OXEN

STOREY'S WORKING ANIMALS

OXEN

★ A TEAMSTER'S GUIDE ★

Drew Conroy

ILLUSTRATED BY BETHANY A. CASKEY

Storey Publishing

The mission of Storey Publishing is to serve our customers by publishing practical information that encourages personal independence in harmony with the environment.

Edited by Deborah Burns

Art direction by Alethea Morrison and Cynthia McFarland
Text design by Vicky Vaughn Design
Text production by Kristy MacWilliams
Cover design by Alethea Morrison

Cover and chapter opener illustrations by John MacDonald
Letterpress typography and borders by Yee Haw Industries
Illustrations by Bethany A. Caskey
Back cover photographs by © Drew Conroy: top; © Janet Conroy: bottom; and © Adam Mastoon: middle
Interior photography credits appear on page 291

Indexed by Susan Olason, Indexes and Knowledge Maps

Printed in the United States by Versa Press
10 9 8 7 6 5 4 3 2 1

LIBRARY OF CONGRESS CATALOGING-IN-PUBLICATION DATA

Conroy, Drew.
    Oxen : a teamster's guide / by Drew Conroy. — 2nd ed.
      p. cm. — (Storey's working animals)
    First ed. published: Gainesboro, TN : Rural Heritage, 1999.
    Includes bibliographical references and index.
    ISBN 978-1-58017-692-7 (pbk. : alk. paper)
    ISBN 978-1-58017-693-4 (hardcover : alk. paper)
    1. Ox driving. 2. Oxen. I. Title. II. Series.
SF209.5.C66 2008
636.2'12—dc22
                            2007042267

Contents

Acknowledgments

This book is the result of years of work with oxen and research into how they were used in the past and are used today. I owe a tremendous amount of credit to my parents Bernard and Kathy Conroy, who allowed me to follow my dream. They never said no when it came to my interests in oxen. They allowed me to raise and train as many oxen as I could afford to feed, despite the many financial struggles they faced. Their physical and moral support were essential in the development of my skills and interest in oxen, and in my eventually putting it all together in this book.

Many other adults guided me through my early years. My first mentors in training oxen were Chester "Punky" Rhodenizer, Rob Gordon, and Jerry Courser. Without their ideas and input I would never have trained my first team.

Showing and pulling oxen as a kid, I learned a great deal from the many 4-H youth and adults who shared my passion. There are too many to name, yet all who shared with me the challenge of competition helped shape my understanding of how to work and train oxen. Many of them to this day continue to provide me with endless ideas, photographic opportunities, and inspiration. These include people like Brian and Kim Patten, Phil Mock, Bob Mock, Dwayne and Faith Anderson, Frank, Pauline and Arthur Scruton, Tim Huppe and family, Les Barden, and the many 4-H club members in New Hampshire who continue to keep the tradition alive.

I owe a great deal to two of my college professors. Dr. Doug Butler, then a faculty member at Northwest Missouri State University, inspired me in 1982 to put down on paper what I knew about oxen. Later, when I was a student at the University of New Hampshire, Professor Dwight Barney pushed me along to finish what I had started in Missouri. Without the intervention of these two men, my interest in oxen, like many childhood hobbies, would have gone no further. Their gentle persua-

sion resulted in an event that changed my life: the publication of *The Oxen Handbook* in 1986.

I owe a great deal to Richard "Dick" Roosenberg at Tillers International. Over the last 15 years his ideas and feedback have been critical to making me rethink what I knew about oxen when I wrote *The Oxen Handbook.* The opportunities he provided me with and the people he introduced me to have shaped my understanding of how oxen are yoked, trained, and used around the world. Dick helped me realize that I could take my interest in oxen to new levels. He and Tillers International have provided opportunities for experimentation and photography that would otherwise have been impossible for me to acquire.

Finally, I owe my wife Janet a lifetime of thanks and certainly more credit than anyone else for helping to further my knowledge about oxen. She put up with the hours, days, weeks, and even months that I was away from home learning about oxen in order to write this new book. Her endless patience and willingness to follow me to ox pulls, ox shows, 4-H meetings, movie shoots, and ox teamster training programs are a mere fraction of what she contributed to this book. My absences while working with oxen in Africa were particularly difficult, yet she never once said anything negative about my pursuit or about the fact that she was the one left doing the farm chores.

She and I have truly worked together in the making of this book. It could never have been written without her. During the last 20 years we raised and trained almost as many teams of oxen. These became the examples that allowed me to make this a work not only of ideas, but also of experiences. Janet's energy, encouragement, proofreading, and ideas have been my inspiration. For this I would like to dedicate this book to Janet Alma Conroy, my wife, my partner, my lover, and my friend.

FOREWORD TO THE SECOND EDITION

MOST YOUNG PEOPLE grow up having a hero of some sort. That hero may be an athlete, a coach, a scout leader, a teacher. Drew and I, like so many growing up in ox-rich New England, found our heroes among the accomplished men who drove oxen, the teamsters. They worked cattle on their farms and in the woods, and come spring they began exercising and conditioning their teams in preparation for the pulling and showing competitions held at the country fairs.

As members of the 4-H Working Steer Program, we trained our own teams under their watchful eyes. Sitting on a hay bale in an old barn or at a fair, we listened to the stories they told and the sound advice they gave, and we strove to take our teams to a higher level of training.

The author and Tim Huppe

Standing among the teamsters during a competition class, in which teams of steers and oxen pulled many thousands of pounds, we heard detailed discussions of proper yoking, yoke design, draft, and the principles of good teamstering. This vibrant oral tradition, however, while passing knowledge on through generations of teamsters, has never been fully realized in print. This book is an attempt to gather those voices on paper.

Drew has dedicated much of his life to this task. His formal training and subsequent position of Professor of Animal Science, the knowledge he has attained from good ox teamsters, and his many years of hands-on experience using his own teams for practical work — all have qualified him to teach both the basics and the advanced craft of training teams and teamsters. He has written dozens of articles, published two books, and made two training films. He has done research in seven countries on four continents. He has worked with and lectured to 4-H clubs, rare-breed organizations, and historical groups, and has somehow found time to conduct in-depth and intensive workshops.

For good reason, Drew is often referred to as an expert on the subject of oxen. I once heard him say, "I don't think of myself as an expert; however, I consider myself to be the best student on the subject of training and working oxen." There is, perhaps, no finer way in which to carry on the work of preserving the integrity of the past while allowing us to touch it in the present. It is his gift to all.

— **Tim Huppe**
BerryBrook Farm
Farmington, New Hampshire

FOREWORD TO THE FIRST EDITION

IN SEARCH OF EVERY CLUE to the better training of oxen, I had carefully studied Drew Conroy's *The Oxen Handbook* several years before I was invited to Plimoth Plantation in 1990 to help with a session on training oxen. At that time I had been working oxen for nearly twenty years.

I knew Drew Conroy would be there with a team. I knew he was still a young man, a product of New England's excellent and much coveted 4-H Working Steer program. I had never met Drew. To spot him I watched each teamster bring his team up the lane.

One team stood out for their sharp responsiveness. The force of the teamster's confidence in working his animals immediately caught my attention. His oxen reacted quickly to every command, although they did not appear to be frightened. His commands were clear and unwavering. Of course that young man turned out to be Drew. My own ox driving has improved significantly as I have studied him and his methods of working cattle. What makes this guy click so well with teams of cattle?

Drew's enthusiasm for working oxen has the same irresistible strength as his authority over them. It is not surprising that the students at the University of New Hampshire honored him for excellence in teaching. His excitement is contagious. He frequently comes to Tillers training center in Scotts, Michigan, to help teach classes in ox training and driving. Drew challenges students with one training exercise after another. He uses stories and slides to stretch the brains of students after their legs will not chase after another steer.

The first time Drew came to teach one of our classes in 1992 he must have spent most of his nights reading to satisfy his driving curiosity. Our library of international literature on oxen attracted him like a moth to a night-light. He would be reading quotes to students next morning at breakfast. He was fascinated by the similarities with which people handled ox problems, even though they had never compared notes or seen each other's work.

But intellectual curiosity alone does not make a great ox driver. Commanding a team of oxen twenty times your weight without lines or leveraged bits is not just an intellectual exercise. An ox driver must be extremely attentive and observant of his/her team. But the teamster must also have will power and determination backing every action and cue to convince the ox that he is following a worthy leader and has no need to question or resist commands. Teachers of draft animal driving, such as Les Barden and Lynn Miller, emphasize this point in their focus on first training the teamster.

Drew, it seems, was born with this requisite determination as part of his personality. Perhaps he has the same self-doubts we all suffer from, but he hides that human frailty. I assure you, no ox sees it in him. I have seen him repeatedly convince problem teams around the United States and in Africa of his superior determination. Verlyn Klinkenborg, after researching oxen at Drew's side for several days while writing "If It Weren't for Oxen We Wouldn't Be Here" for the August 1993 *Smithsonian*, skillfully described Drew as "purged of uncertainty."

This determination in Drew's character is, I am convinced, a key to his success with ox training. While you may not have inherited it from your parents, Drew will teach you its utility. He will urge you to never let a young ox run away from you, even though you have no lead rope or line on him. You will ask, "How do I do that from the side of the team?" You will certainly have moments to wonder as you take on your first team.

If you have the opportunity to watch Drew, you will see him reach out his arm and goad a half-mile to flick a team on the nose to remind them that they just ignored his command to stop. If they are beyond the long arm of his goad, he will project his voice three-quarters of a mile to bring them to a stop. If they should be so foolish as to ignore that, he will pretend that he let them run ahead until they are about to hit the next tree or the side of a barn. Then he will cleverly command them to stop just before the barn or tree slams into their foreheads. Drew urges ox drivers to use every bit of the environment for the education of the ox.

Drew would not be nearly the ox driver he is without his deep appreciation of cattle. He must know the names of half the cattle in New England. He immediately spots each animal's many traits. He knows the boundaries within which cattle feel comfortable. He provides repeated reinforcement of positive behavior. I suspect that when a nine-year-old Drew was training his first team late into the night through all the lanes in the forest, he got as much satisfaction from an approving lick to his hand after a long venture as his oxen gathered from his approving, "Good boys."

Mark Twain once quipped, "He who grabs a bull by the horns, learns twice as fast as he who reads a book." Although Drew draws on his university resources to focus light on aspects of ox care, you can be certain this author has been where he asks you to walk. Drew speaks of oxen from exceptional personal experience. It is our great fortune that he has had the patience to put it down on paper.

— **Dick Roosenberg**
Tillers International

PREFACE

ON A COLD NIGHT IN 1978, my dad drove me 30 miles over snowy New England roads to look at a set of Brown Swiss twins. It was love at first sight. For $50 I bought the calves from Donald Hawes of Milford, New Hampshire. Being young, excited, and impatient to embark on my new adventure with my first pair of steers, I talked my dad into taking them home that night.

We loaded the calves onto the back seat of our two-door Toyota Corolla. Of course the young steers made a mess of the seats, and I scrubbed and cleaned the car before my mother saw it. As I look back on those early days, I can't believe the things my parents endured for my love of working steers and oxen.

Once the calves were home, I faced what seemed like daunting challenges. The calves came down with scours. They were reluctant to follow me on the lead rope and even more reluctant in the yoke. I broke my leg two months after purchasing the calves. Neither of my parents had raised or trained oxen, so they could offer little advice.

Somehow I managed to train Zeb and Luke. I suppose I was motivated by my father's insistence that we could always eat them if I couldn't get them trained. I matured a lot while training those steers. One of the greatest lessons I learned was that there is no substitute for time spent working with animals. I also learned the importance of communicating with people and learning from the experience of others.

I learned how to train oxen from local experts. These were the men who had maintained the cultural tradition of driving oxen in New England, when the rest of the country had largely abandoned these animals nearly a century earlier.

I wrote *The Oxen Handbook* in 1985 after I realized how difficult it was to find information on raising and training oxen. I was constantly searching for information on oxen as understanding how to work and train my steers became an obsession. Being very shy (my friends today don't believe that) I would rather have read books than asked strangers questions about training oxen. My quest for information forced me to overcome my shyness.

As a boy I never dreamed my oxen or my ideas would be featured in a number of films, videotapes, and magazines. Nor did I think my advice would ever be sought by people using oxen in competition, in agriculture, and in the forest. To this day I continue to seek the advice of many ox teamsters with more experience than myself. Attending oxen pulls and shows with my own teams subjects me to the criticism of the local experts whose knowledge and experience I value highly.

This book emphasizes New England methods of training and working oxen. Keeping ox teams today is primarily a hobby, and most oxen are used for exhibition and competition. Hobby oxen in this country are certainly less important than oxen used by farmers around the globe who depend on them for food production. Oxen, however, have been used in the forest and fields of New England for centuries, much as they are still used in other areas of the world today.

A team of oxen is the result of human-directed training and thought. For centuries many New England ox teamsters have begun training their animals as calves. Although the practice of training calves is not universal, it creates animals that are easy to work with and eager to please. Starting with calves is recommended for beginning teamsters and children interested in training steers or oxen.

For centuries, New England children have trained oxen. They often command such high levels of performance from their animals that they become the envy of many adult teamsters. Dozens of regional competitions see scores of children compete, mesmerizing crowds with their animal-handling skills.

Ox pulls and shows have always been a mainstay at agricultural fairs in the northeastern United States. These competitions create an environment where ox teamsters are quick to adopt new ideas and hone techniques in training and working cattle in order to gain an edge in competition. Much like other livestock competitors, ox teamsters in the United States strive for perfection in selecting, training, yoking, and working their teams. This opportunity has helped New England ox teamsters maintain a system of yoking, training, and driving oxen that is a unique New England phenomenon.

The ox was an important beast, and his image was recorded in many early photographs. Old yokes may still be found hanging in New England buildings. Were it not for the poor farmers who kept oxen in the yoke, the skills and techniques needed to train and drive them would have likely been lost.

I wrote *The Oxen Handbook* as an undergraduate documenting the tradition of driving and training oxen in the northeastern United States with little regard to their use or importance in other nations of the world. Since writing that book, I have gained a great deal of respect for other systems of working and training cattle. This respect and admiration come from personal experience with oxen and ox teamsters all over the United States, as well as the countries of Tanzania, Uganda, Kenya, Canada, England, and Cuba. Furthermore, my travels and communications have put me in touch with farmers, prominent researchers, authors, extension agents, and museum personnel from around the globe who promote and use oxen in their work.

As my own children grow and develop their interests in things other than oxen, I have found less time to work with my own teams. Conducting ox training workshops and working with new teamsters inspire me to get my team out of the pasture. I continue to make time for the youth in the New England 4-H Working Steer Program, where I got my own start with oxen. These hands-on educational activities have kept me in touch with the challenges that new ox teamsters face. They also remind me of the constant challenges of keeping a team comfortable in the yoke and willing to do any task demanded of them.

This book still emphasizes my cultural roots of learning to train and work cattle in New England; however, the principles can be applied anywhere. It has been amazing to observe other cultures outside the United States who use the same basic principles I have outlined in this book to train their animals. The challenge for new ox teamsters the world over is that cattle have no instinct to work. If you want them to respond to you in the yoke you have to follow through with a sequence of understanding the animals, then handling, and finally training, as outlined in chapters 5 to 7.

Anyone with patience, a basic understanding of cattle, and a sincere commitment to spend some time and effort can successfully train oxen.

1
SELECTING
THE
IDEAL TEAM

hether you are selecting your first or your fiftieth team of steers to train as oxen, your preference as to breed and type, as well as your plans for how the animals will be worked, will and should influence your choice. Although your preferences are important, you would be wise if you are a novice to seek the advice of experienced teamsters when choosing one steer over another. You want the time and money you invest in these steers to result in pride, rather than disappointment.

What Is an Ox?

An ox is not a separate species or breed, but simply a castrated bull used for work. While many examples from around the world display or describe both cows and bulls being used as draft animals, castrated male cattle are most often employed as work animals.

The castration of a bull at an early age changes his growth pattern. A growing steer will develop longer legs and a larger frame. Thus the ox grows taller and heavier than the bull of the same breed when he is castrated before he matures. With castration and the loss of the influence of the hormone testosterone, the steer is also more placid and more easily controlled than the bull. Although castration causes a loss of muscle mass, particularly in the neck and shoulders, the greater size, length of leg, and more agreeable attitude of the ox make him a steadier and more desirable work animal. Castration, particularly on a small calf, also influences horn size and head shape, giving the ox a more cow-like appearance.

The two existing species of cattle, both used for work, belong to the genus *Bos*. They are *Bos indicus* (the Zebu, or humped breeds) and *Bos taurus*, representing breeds of European or Asian descent that have no humps. *Bos indicus* is common in warmer climates, while *Bos taurus* is more common in temperate and cool climates. For use as oxen, the two species are often crossbred by farmers in some areas to combine traits both species possess.

The ideal ox is large, strong, straight, and healthy.

A third, now extinct, species was the *Bos primigenius* or auroch (also called urus), a huge wild animal thought to be similar in size to the modern Italian Chianina. The domestication of such an animal represents a great milestone in human history. The auroch is thought to have been a predecessor of, or at least had a great genetic influence on, early domesticated cattle. The auroch became extinct during the early seventeenth century. The last known survivor died in captivity in Eastern Europe.

Pulling Together

Oxen are normally worked in teams of two, with pairs sometimes hitched together for additional power. Multiple pairs form teams used for logging, heavy transport, and plowing dense soils. Teams are hitched together in tandem, one pair after the other. Out of necessity, often where farmers have few resources, oxen have been hitched together with other species, such as the horse or donkey. Even the poorest farmers, however, admit that multiple-species hitches are not a desirable situation.

Selecting a Team

Selecting steers to train as oxen is much like selecting any animals to be used for work or production. Before choosing an animal you must have a standard by which to evaluate the animal. No set standard has been established for conformation of the ideal ox, but most teamsters agree that an ox should be healthy, well muscled, and sound on his feet and legs, and have an agreeable disposition. A well-matched team of carefully selected and trained animals, with good conformation, is a joy to work. Such animals can be expected to live for many years and may be used for a variety of tasks on a farm or in the forest.

Oxen are generally easier to train and work as a pair or team. Selecting a team, however, is more challenging than finding one good animal. Putting together a team involves finding two or more animals that match not only in color but, more importantly, in disposition, agility, and size. Few perfect teams exist. Minor differences between animals are hard to avoid. Teamsters do the best job they can of matching or "mating" (putting together or breeding cows to bear calves that will grow, develop, and act alike) their teams.

THE IDEAL OX

An ideal mature ox may never be seen or found, but it is always something to strive for in selecting animals. Understanding conformation and knowing what to evaluate is an important part of the selection process. Great differences exist between breeds and types of cattle. Take these differences into account in evaluating your animals for work. A breed like the N'Dama, a tiny West African animal, will have a different standard size and temperament than the giant Chianina, developed in Italy under much different conditions.

Before trying to put together a team, have a mental picture of what an ideal ox looks like. A good judge of livestock starts with a breed standard or scorecard. As no set standard has been developed for judging or evaluating the ox, you have to learn what to look for. Most experienced teamsters agree that an ox should be able to perform the duties expected of him for many years with a minimal amount of difficulty or hardship. Certain characteristics of the body are therefore essential. (See box, next page.)

SHOW, PULLING, OR FARM WORK?

As an ox teamster, have specific goals in mind for your team. How will you use the animals? Will you use the team for show, pulling, or farm work? Suitability to the type of work will greatly affect your choice of animals.

Pulling oxen and oxen used for farm work alone do not have to match in color or horn shape. The focus is work; therefore, the temperament and the ability of the two animals to work together effectively under a heavy load is more important than matching perfectly as a team.

A show team should be calm and obedient, but also well matched and flashy with an appearance

What to Look For in an Ox

The ideal ox is large for his breed, and he stands on a strong and relatively straight set of front and rear legs when viewed from the front, rear or side. The animal is not too cow-hocked, post-legged, or sickle-hocked. His hooves are dark, short in the toe, high in the heel and wide, but not to the extreme. His feet and legs track forward in a straight line. His neck, shoulders, back, rump, and thighs are heavily muscled. Working cattle transfer most of their energy from their rear legs, which push forward, to the shoulders or neck and head, depending on the type of yoke. To easily transfer this energy, the body, chest, and rump must be wide and strong. Every animal used for work should show the desirable degree of thickness for his breed.

Other essential traits include the following:

- Freedom from disease and conformational defects, such as a weak back, poor teeth, or any other defect of the mouth.

- A clean healthy haircoat.

- Youthfulness in age and appearance.

- Clear bright eyes, free of cloudiness or lesions.

- Alertness and ability to hear without any difficulty.

- A temperament that allows him to respond to normal training.

- When a head yoke will be used, large solid horns.

Pass over steers that lack any of these essential characteristics. You will spend a great amount of time and energy training your team, and training inferior animals does not often yield desirable oxen. Characteristics such as color, breed, and how well the animals match are much less important than their ability to perform in the yoke.

While you might argue that animals lacking some of the listed characteristics could make fine work animals, good conformation helps ensure that your animals will not develop problems during their years of hard work. Animals lacking perfect conformation may work successfully, but poor conformation or health problems eventually catch up with them.

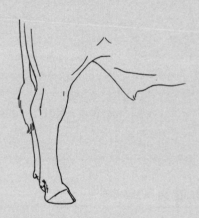

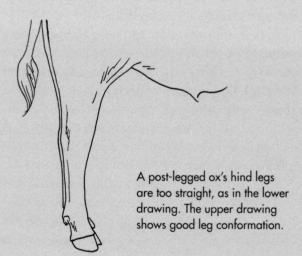

A post-legged ox's hind legs are too straight, as in the lower drawing. The upper drawing shows good leg conformation.

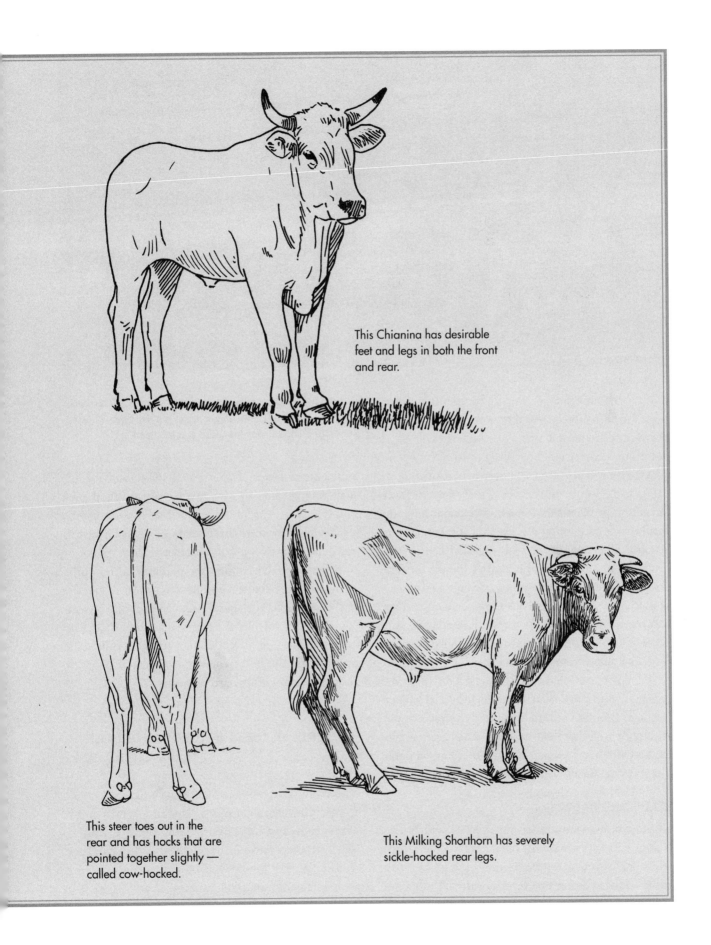

This Chianina has desirable feet and legs in both the front and rear.

This steer toes out in the rear and has hocks that are pointed together slightly — called cow-hocked.

This Milking Shorthorn has severely sickle-hocked rear legs.

that catches the judge's eye in competition. Most oxen used for show are well matched in body, head, horns, and bone structure. They often carry more flesh than teams used for pulling or work.

A pulling team used in competition or logging needs to be active in the yoke, free of major conformation problems and, most important, have the mentality and willingness to pull aggressively as a team. They should be well muscled in the legs, neck, back, and chest. Their legs should be relatively straight and their top lines straight and strong. Due to their appearance or character, many of the best ox teams in pulling contests or woodlots would never place high in the show ring. A hard-working ox doesn't have to be attractive.

A farm team need not look sharp or have the willingness to push their physical limits. However, the basic selection criteria for sound oxen still apply. Although well-matched animals won't necessarily work any better for you, they will be more valuable if you decide to sell them.

SELECTING CALVES

The beginning teamster wishing to work with horses is advised to start with an older, well-trained draft team. By contrast, the beginning ox teamster can easily train his or her first team of calves. Both team

A farm team need not be flashy like a show team, nor pull aggressively like a competition team, but must be willing to work, and should match in case you sell them.

and teamster learn as they go. New England 4-Hers as young as eight or nine years old train their own calf teams with tremendous success. A well-trained team that is two or three years old can be just as calm and easy to work as an older team.

Selecting calves allows the teamster to mold the animals' sizes, shapes, and personalities. While the ability to mold the team is not foolproof, a teamster with control over the team's feed, handling, and environment will have a greater impact on how the animals mature in the yoke than a teamster who lets his calves run with a herd of cattle.

A team of oxen is made, not born.

A team of oxen is made, not born. It is the result of many hours of handling, training, and careful attention to feed and environment. Some teamsters buy four calves and start two teams, so they can later choose the best two animals for their team, selling the two animals that don't match or work as well.

SELECTING MATURE OR TRAINED ANIMALS

Some teamsters prefer to let someone else do the raising, matching, and training, then choose the best available animals to meet their needs. This method is more foolproof than buying calves, because older animals are less likely to change in personality or size. The challenge is that oxen are not as common as they once were. In some countries where draft cattle are more common, the only animals sold are culls. Most good stockmen never sell their best animals.

Oxen live only 10 to 15 years. As an ox teamster once said, ". . . just about the time you get the team broke just right, they up and die on you." Many teamsters who have good teams and use them daily do not cull their animals. You may therefore have a hard time finding a mature, well-trained team for sale at a reasonable price.

Selecting untrained animals at maturity is a good way to know exactly what kind of animals you are getting. Training them will be more challenging, but you do not have to wait for them to grow up to be put to work, nor must you wait to see what they will look like. I have purchased calves that matched perfectly, only to have the two animals grow up to be different colors with different conformations and dispositions. Eventually I had to buy new mates for each of the already-trained steers.

Choosing a team as mature or older animals is typical in many developing countries, where cattle are not large and they are handled daily while being herded to and from grazing areas. Teams are put together and almost immediately trained and put to work. The system works well because the animals are accustomed to people and are small enough to be muscled around, if necessary. The system is not as effective in the United States and Europe, where cattle can weigh more than a ton and may never have been touched prior to training.

Breeds

Whenever breeds of cattle are discussed for use as oxen, debate and differences of opinion ensue. With so many breeds of cattle to choose from, opinions are endless. It is clear that no one breed is best for all purposes. My own breed of choice has changed several times over the years as my level of experience, age, and needs have changed.

For example, I admire and have trained numerous Devon teams for their speed, stamina, and attitude, but other breeds are calmer and easier to work with. The Devon breed has proved to be too much for my children, who want to work casually with my team. The Devon has also proved to be difficult for other novice teamsters, in ox-training workshops.

An important question to ask early in the selection process is whether or not the breed you desire is available in your region. You might choose a breed fitting all your needs, but the nearest herd or team is 2,000 miles away. Are you willing to spend the money or time necessary to acquire such a team? Is the importation of an exotic breed really the best choice? Often more common breeds or local strains are better adapted to local environments and diseases, an especially important consideration in developing countries.

Criteria for Selection

When choosing calves, select animals from similar bloodlines. They are more likely to remain matched when mature. Important factors in selecting a team include:

Disposition

Disposition involves an animal's attitude toward being in a yoke and following commands. Some cattle are easy and placid in the yoke; others are high-strung and need constant attention. In selecting a team, make sure you have animals that work well together. Slight differences in disposition that need to be worked out always exist, but an aggressive or nasty steer yoked with one that is friendly and calm is not a pleasant experience, especially if the steers constantly fight each other in the way they move in the yoke.

Occasionally a well-trained steer is used effectively in training a wild one. Historically, this practice was common in logging camps and freight companies. Today most oxen in the United States have been trained from an early age.

Agility

Some cattle are more alert and spirited than others. Don't yoke two animals that are opposite in agility. A fast steer and a slow one won't make a good team for show, farmwork, or pulling. In both cases one steer is always ahead of the other, limiting the team's effectiveness. Ox teamsters around the world struggle with two animals that are not matched in their willingness to work as a team.

Size

Two steers that are different in size may work effectively in the yoke, provided they have a similar disposition and agility level. Similar size is important for a show team, as differences may affect the judge's placing. Keeping two steers exactly the same size by altering their feeding program can prove difficult, especially if they are purchased as calves.

While modern ox yokes can accommodate a small steer and a large one, a great deal of experience is needed to make and adjust a yoke to deal with animals of different sizes.

Conformation

Many ox shows and judges consider how well a team matches and the animals' overall conformation. Slight differences in conformation are not usually a problem in a working team, but all other things being equal, an unevenly matched show team will not place as well. A team used for show and exhibition should be matched in body conformation. You can fairly easily put one animal on a diet to maintain his weight, but you will have a harder time changing his frame and overall size.

The way the team walks, stands, and moves is sometimes a function of conformation. It is not critical that the animals match perfectly in conformation, but vast differences can affect the team's appearance and sometimes overall value. Many experienced teamsters strive for heavy-boned, lean cattle with large feet and a steep hoof angle (short toes and high heels) because such animals supposedly live longer and have fewer health problems.

Color

Often one of the most sought-after characteristics in selecting cattle, color may give you some indication of the animals' breed and bloodlines, but it remains one of the least important traits for a working team. A team that is well matched in color is attractive, but if the animals are not similar in disposition, agility, size, and conformation, problems arise as the animals are put to work.

THE EFFECT OF SPECIALIZATION

In developed countries, many breeds of cattle origi-nally selected for size and strength in the yoke have been intensively bred for meat or milk production. Where once the speed, working ability, and adapt-ability of cattle were valued as desirable character-istics, modern breeds have been selected to fit neatly into categories and systems of intensive modern cat-tle management.

Beef cattle have been selected for maximum growth and the ability to fatten quickly and easily, while other attributes important to an animal used for work have been neglected. Such characteristics often include sound feet and legs, athletic ability, and tractability. Beef calves are hard to find in the north-eastern United States. They may be more expensive to purchase and more difficult to train than dairy breeds, largely because of their lack of human con-tact and regular handling at a young age. The beef breeds are thick and heavily muscled, and as oxen they may be difficult to keep lean for pulling com-petitions based on weight. In Nova Scotia, however, beef breeds such as the Hereford and its crosses are commonly used. Many of these fine animals are brought to New England to compete in pulling con-tests and easily beat leaner dairy-type cattle.

Dairy cattle have experienced a greater degree of specialization than have the beef breeds. Short-horns, Brown Swiss, and Ayrshires were once bred as triple-purpose animals to provide work, meat, and milk. Many of the historical qualities like har-diness, muscling, and soundness have been for-feited in the name of high milk production. This fault lies not with the breeds or breeders, but with their attempt to survive in a world of specialization and maximum efficiency. As oxen, dairy breeds and their crosses are generally more popular than beef breeds in New England, largely because of tradi-tion. Compared to beef breeds, dairy calves may be purchased younger, trained more easily, and molded into exactly what the teamster wants

Dual-purpose breeds have retained more of the original characteristics desired by ox teamsters, but in many regions their numbers have fallen rapidly. The American Milking Devon is a classic example. This breed was once one of the most sought-after

Losing the Draft

Today the American Milking Devon is one of the only breeds in the United States in which breed-ers have actively tried to maintain draft charac-teristics. Consequently, this breed is now rare. It cannot compete with the Holstein or Jersey in efficient milk production, nor can it compete with its close cousin the Beef Devon or with the Cha-rolais in the feedlot, where the only measurement of success is a larger, faster-growing, and more heavily muscled animal.

animals in New England. In 1880 J. Russell Man-ning, a veterinarian and author, wrote about the great attributes of the Devon: "As workers, milkers, and beef makers combined, for the amount of food taken in they have no superior."

In the United States and Western Europe the survival of any breed has been based on its ability to adapt to drastically changing markets and meth-ods of management. Many cattle breeds originally began as triple-purpose animals, but the draft char-acteristics that once were common are now difficult if not impossible to find. A smart, active, and ath-letic animal does not pay the bills.

CHOOSING A BREED

More than 50 breeds of cattle are found in the United States and more than 1,000 breeds in the world. While certain breeds may appear similar to one another, their temperament, color, or other characteristics vary. All cattle can be trained to work. All have their own individual personalities, yet breed can greatly influence temperament and speed. Some teams need more work than others in order to maintain their attention and control.

When choosing a breed, consider the type of environment in which the animals will be worked. Some breeds are better suited to certain climates or environments. Choosing a beautiful team that can-not endure the heat, insects, or local diseases will affect the team's working ability.

Animals used in hot climates, for example, need to be suited to such weather. Zebu or *Bos indicus*–type cattle are better suited to warm climates. They will work harder and faster than European or *Bos taurus*–type animals exposed to the same degree of heat and humidity.

The opposite holds true in cold climates. A team used in logging operations in northern areas must be acclimated to such conditions. Many Zebu animals, with their loose skin, short hair, and ability to endure heat will suffer greatly in subzero weather.

In areas where trypanomiasis and other tick- or insect-borne diseases are endemic, certain breeds of cattle are unable to survive. Native breeds with some natural resistance are often preferred in developing countries, even though larger, stronger animals might be available.

Most teamsters have a favorite breed. Others seek out characteristics they claim are just as important as conformation or temperament. Some teamsters never yoke a steer with a black nose or white feet. Although characteristics like skin or hair color may affect an animal's ability to work in direct sun, some of these "rules" have no bearing on an animal's ability to work.

Don't believe everything you hear until you have had the opportunity to see numerous breeds in the yoke. When you find a breed you are interested in, ask the breeder how easily they train to a halter or how well behaved they are when captured. Usually individuals within a breed behave in a similar manner on a halter and in a yoke.

Temperament is a heritable trait passed down the generations. Beware that some lines or families within a breed can be more difficult to work with.

COMMON BREEDS

The most common breeds used as oxen in New England include the Holstein, Milking Shorthorn, Brown Swiss, Chianina, Jersey, and Milking Devon. Also to be found are teams of Ayrshires, Dutch Belted, Charolais, Longhorns, Herefords, and Dexters.

Milking Shorthorns and Holsteins are considered excellent breeds for the beginning teamster because of their even temperament and lower purchase price compared to other breeds. Dairy calves have the additional advantage of being handled often and bottle- or bucket-fed, which makes training easier. Modern dairy breeds tend to be leaner and more angular than those of the past, but the mature ox usually fills out and puts on muscle if given the opportunity. Many experienced teamsters will yoke nothing but a Holstein or Milking Shorthorn.

The unique characteristics of breeds such as Chianina, Dutch Belted, and Milking Devon make them desirable as oxen, but these breeds are hard to find. They are more expensive than more common bull calves that can be purchased from a local dairy farm. Many other breeds may catch your eye but may not necessarily suit your needs as a teamster. Just because you are impressed by a pair of giant Chianinas or a pair of snappy Milking Devons, don't be fooled into believing you will be less satisfied with oxen of a more common breed.

When a certain breed is not available in your area you might convince someone to breed some cows to a bull of that type, if you provide the semen. Shipping semen across country is a lot cheaper than shipping live cattle the same distance. Artificial insemination has produced many crosses without a major financial commitment or risk to the breeder.

HYBRID VIGOR

Many teamsters prefer crossbreeds because they combine the best characteristics desired in a draft animal, including soundness, longevity, and hybrid vigor. These crosses work well since neither milk production nor maximum growth is an important selection factor. Even multiple crosses or "mongrel" cattle make terrific work animals, because they tend to be more like their early predecessors. The challenge in selecting such animals is finding a suitable teammate.

Far too many breeds exist to cover them all here, so I'll discuss only the breeds most commonly used as oxen. Numerous books are available (some listed in this book's appendix) that describe the characteristics of cattle breeds around the world. While the following descriptions offer a general guide in choosing a breed appropriate for your needs as a teamster, hundreds of other breeds are used equally successfully as work animals. When the training is done, any team with good conformation is impressive, no matter what the breed.

Breed Comparison

Every team is the result of hundreds of hours of training and work. Any breed of cattle may be trained. Given an equal number of hours in the yoke, some are easier to work with. They would receive a 1 to 3 on the tractability scale. The more challenging breeds (rated 7 and up) may have traits that are desir-

able for specific purposes, such as pulling competitions. Such breeds are preferred by teamsters who seek their special characteristics and have the skills to handle the animals. A great deal of genetic variability exists among the various breeds. The wide variability in size and temperament means a breed is out there to meet the need of any ox teamster.

See appendix I for a chart comparing ox breeds.

AYRSHIRE

Origin	Scotland
Color	brownish, red, and white
Mature Weight	2,000 lbs. (900 kg.) (average)
Height	medium
Temperament	active and alert
Tractability	7
Other	attractive long horns

Noted for its long horns that grow out, then tilt upward and backward, this medium-sized dairy breed originated in Scotland in the county of Ayr. The mature Ayrshire ox in good working condition matures to nearly 2,000 pounds. Ayrshires are reddish brown and white and can come in any combination of those two colors, including almost all white to brindle.

Like other breeds developed in Scotland, Ayrshire cattle tend to be active and aggressive on pasture, and they are noted for their great ability to thrive on poor pasture. In the yoke they train easily

The Ayrshire has red and white patches and distinctive horns.

as calves, but they may be challenging if allowed to have their freedom for the first few months of life.

BROWN SWISS

Origin	Switzerland
Color	light to dark brown
Mature Weight	2,400 lbs. (1,090 kg.) (average)
Height	large
Temperament	docile and slow moving
Tractability	2
Other	grows rapidly

This large breed grows fast and carries more flesh than other dairy breeds. It was originally bred for size, ruggedness, and sound feet and legs, characteristics that persist today. The Brown Swiss is noted for its docility, slow (sometimes poky) nature, and late maturity.

Teamsters who are accustomed to other breeds may find this breed to be temperamental, sensitive, and difficult to work. Temperament should not be confused, however, with an inability to work. Many Brown Swiss teams are excellent work animals for the field, forest, show ring, or pulling competition. In fact, in many Latin American countries, Brown Swiss cattle are sought after and often crossed with Zebu animals.

The Brown Swiss has unique coloring: solid brown that may vary from light grayish brown to almost black. Bulls are generally dark, and the longer bull calves remain intact, the darker they usually become. They always have a black nose, tongue, tail switch, and hooves, and a unique white ring around the muzzle.

The calves are generally large and grow quickly. Calves are sensitive and may require extra care for the first few weeks, especially if they are bucket-fed and not allowed to nurse from a nipple pail or cow.

In New England, Brown Swiss are used in heavy pulling classes because of their large size. Mature animals weighing up to 2,500 pounds are common. Due to its docile nature and slow pace, this breed is popular among youngsters, although the fast-growing calves soon tower above their young teamsters.

CHAROLAIS

Origin	France
Color	white
Mature Weight	2,300 lbs. (1,045 kg.) (average)
Height	large
Temperament	moderately alert and active
Tractability	6
Other	heavily muscled

Mature animals of this large white beef breed are large framed and heavily muscled, making them fine draft animals. The calves are sometimes hard to find. This was once the draft breed of choice in France, but the number of Charolais cattle used as draft animals in New England is low. The Charolais is often crossbred with the Holstein to get a light brown or tan, almost khaki-colored, animal.

CHIANINA

Origin	Italy
Color	white
Mature Weight	3,000 lbs. (1,365 kg.) (average)
Height	huge
Temperament	alert and excitable, at times to a fault
Tractability	10
Other	grows rapidly

Developed as a large draft animal in the Chiana Valley of central Italy, this is the largest breed of cattle in the world. Mature steers are often 6 feet tall at the withers and weigh up to 3,500 pounds. Beautiful spans of Chianina oxen can still be found in rural Italy today. They are fast growing and especially long legged. Of all the breeds, they have one of the most athletic appearances.

The true Italian Chianina is all white with large horns and a black nose, tail switch, and tongue. The breed is noted for its strong feet and legs and for its longevity.

In the United States, many registered Chianina cattle are black or dark brown and polled. Such animals are American adaptations, largely the result of an open breed herdbook. The goal in crossbreeding was to yield more desirable carcass and feedlot traits. This has created an animal different from the Italian Chianina, originally developed to be one of the world's largest, most athletic, and powerful breeds.

The Chianina is popular among teamsters in New England interested in entering pulling competitions. These animals tower above their Holstein, Shorthorn, and Brown Swiss counterparts and at maturity usually win pulling competitions because of their sheer size. In weight classes they also have a higher ratio of muscle and bone compared to gut. In fact, their shape has been compared to a greyhound dog, allowing them to be larger animals in the same weight class.

Since they tend to be more high-strung and excitable than the Holstein, Shorthorn, or Brown Swiss, take care to properly train these animals when they are calves. This extra care is especially necessary when training calves that nurse their dam and are given free run of the pasture. Frequent tying and handling of the calves will pay great dividends.

For use as oxen, the Chianina is commonly crossed with the Holstein, Brown Swiss, or Shorthorn. Most often the Chianina is used as the sire, crossed with large dairy cows to get calves with the slower growth pattern and low-key temperament of their dam. Compared to full-blooded Chianinas, such crosses are competitive in pulling contests, with a calmer temperament.

DEVON

Origin	England; United States
Color	red
Mature Weight	1,700 lbs. (770 kg.) (average)
Height	medium
Temperament	quick and alert
Tractability	10
Other	easy to color match

Beef Devons and a small population of Milking Devons both exist in the United States. According to New England ox teamsters, "The breeders of Beef Devon have bred the brains and brawn right out of their animals." Most Americans who seek Devons for work will not consider the Beef Devon.

The Brown Swiss is a large, calm, slower-moving breed.

The Charolais is a large breed that has been used for centuries as oxen in France.

The Chianina is the largest breed of cattle in the United States. The all-white, horned cattle are the original Italian type.

The Milking Devon is a rare breed. Only about 500 cows survive in the United States. Supposedly descendants of the original cattle used by early pioneers, they seem to have retained most of their original characteristics. The Milking Devon closely resembles early cattle from North Devonshire in England, where it was one of the fastest, most agile, and adaptable of all the English breeds in the yoke. The Devon is small in size, but what it lacks in size, it makes up in character.

The Devon is all red, varying in shade from a deep rich red to light red. Devons have yellow skin and creamy white horns with black tips, and usually a whitish gray tail switch.

The Devon is a fast-moving breed, with a long history as a superior draft animal.

Dexter cattle are one of the smallest breeds of cattle in the United States. They are quite lively.

DEXTER

Origin	Ireland
Color	black, dark red, dun
Mature Weight	1,000 lbs. (455 kg.) (average)
Height	small
Temperament	quick and alert, at times to a fault
Tractability	9
Other	long horns, heavy muscling

This small all-black or dun breed, developed in Ireland, has gained popularity in the United States among small-scale farmers. Dexter working steers or oxen can be found at many ox competitions. They are a nice size for people concerned that cattle may get too large for their facilities, transportation options, or children.

Dexters tend to be high-strung, however, often a function of being allowed to nurse instead of being bottle-fed. They are heavily muscled for their size and grow long horns. They are great for teamsters interested in keeping the animals for fun and to do some light work around the farm or woodlot. This is a good breed for experienced younger teamsters interested in a team that won't outgrow them.

DUTCH BELTED

Origin	Holland
Color	white belt around black body
Mature Weight	1,700 lbs. (900 kg.) (average)
Height	medium
Temperament	moderately docile
Tractability	3
Other	rare breed

Called Lakenvelders or Lakenfeld Cattle in their native Netherlands, this breed was developed as a dairy animal and should not be confused with the polled Belted Galloway.

Although considered rare, Dutch Belted are popular as oxen. They are easily distinguished by their color, which is all black except for a continuous white belt around the barrel that begins at the shoulders and extends almost to the hips.

A dairy breed that is lean and angular like a Holstein, it is of medium size and grows in pace and size similar to the Ayrshire. Mature steers in working condition weigh about 1,600 to 1,800 pounds. In temperament they are similar to the Holstein — generally easy going and calm, but not as docile and poky as the Brown Swiss or Hereford.

HEREFORD

Origin	England
Color	dark red with white face
Mature Weight	2,200 lbs. (1,000 kg.) (average)
Height	medium
Temperament	moderately docile
Tractability	3
Other	rare breed

The Dutch Belted is a striking breed, medium size and easy to match together.

Originally developed as a draft animal in England, the Hereford is reddish brown with a white face and belly, often with a narrow white stripe along the top line. The breed is medium to large, fattens easily, and tends to carry more flesh than it did in the old days. Individuals are easy to match in color, but can vary greatly in size. They are easygoing and calm.

In the United States, Herefords have changed considerably over the past 150 years. Originally used in the eastern United States as multipurpose animals that met all the needs of the small farmer, they were later selected only for beef. Their value as draft animals dropped tremendously when they were bred to be short and fat at the turn of the last century. Modern animals are large, heavily muscled, fast-growing, and efficient beef animals.

The most popular type today is the Polled Hereford, meaning it has no horns. Horned Herefords are still around and occasionally seen in a yoke. In Nova Scotia, the Horned Hereford is a preferred breed to cross with Shorthorns or Ayrshires to produce attractive, heavily muscled animals to use in the head yoke. Hereford-Holstein crosses are also popular among today's New England ox teamsters.

The Horned Hereford is not as common a beef animal as it once was, but its calm nature is a trait the breed is known for.

HOLSTEIN

Origin	Netherlands
Color	black and white or red and white
Mature Weight	2,500 lbs. (1,135 kg.) (average)
Height	large
Temperament	docile
Tractability	3
Other	grows rapidly

Dutch settlers first brought the Holstein to North America from the Netherlands in 1621, but the breed was not kept pure. Later importations during the 1850s and 1860s were more successful. These cattle are often referred to as Friesians in other countries, and the Beef Friesian is found as a breed in the United States.

A large breed, the Holstein is noted for its rapid growth with little fat accumulation until maturity. A two-year-old animal may weigh more than 1,400 pounds. Mature animals often weigh well over a ton, with some reaching 3,000 pounds. This breed is not as tall as the Chianina, but some animals reach almost 6 feet at the withers.

The Holstein is one of the more commonly used breeds for oxen in New England. It is popular for beginning teamsters because the calves are easy to find and relatively inexpensive to purchase. While Holsteins are readily trained, however, they often grow too quickly for a small teamster.

Matching a team is challenging because of the great variation in color patterns. Holsteins are normally black and white, but occasionally red and white animals can be found.

Easy to find, the Holstein (top) is a popular breed of dairy cattle that make fine work animals.

The Jersey (middle) is a smart and active breed that can make fine work animals for the farm or show ring.

These oxen (bottom) are Randall-type Linebacks, a medium-sized breed with a striking pattern.

JERSEY

Origin	England
Color	any shade of brown
Mature Weight	1,600 lbs. (725 kg.) (average)
Height	medium
Temperament	moderately active and excitable
Tractability	7
Other	easy to color-match

This small breed of dairy cattle was developed on the island of the same name in the English Channel and was first brought to the United States in the early 1800s. Jerseys are fine boned and lack the ruggedness of the heavier breeds. They tend to mature early, which can be an advantage in pulling competitions among lightweight animals. Well-fed mature steers weigh about 1,600 pounds. Jerseys vary from light brown to almost black, with or without white patches.

This active, alert animal tolerates warm climates better than other dairy breeds found in the United States, but its small size and light muscling can be a disadvantage for heavy work. Jerseys are commonly trained by young teamsters and make adequate work animals for the farm.

LINEBACK

Origin	England
Color	black with white stripe down back and belly
Mature Weight	2,000 lbs. (900 kg.) (average)
Height	medium
Temperament	moderately active and alert
Tractability	6
Other	rare breed

In the United States Linebacks are varied in genetic background, yet black and white animals similar to the Gloucester cattle of England can still be found as oxen. The Randall Lineback, white with blue roan coloring, is also found in the United States.

The Lineback cattle are popular among people who show oxen because of their striking appearance — black or blue roan with a white stripe along the back and belly — and because they are active but agreeable in the yoke. Linebacks are of medium size, with mature steers weighing from 1,600 to 1,800 pounds.

MILKING SHORTHORN

Origin	England
Color	red, red and white, roan
Mature Weight	2,300 lbs. (1,045 kg.) (average)
Height	large
Temperament	moderately docile
Tractability	4
Other	matures slowly

One hundred years ago the Milking Shorthorn was the most sought-after breed, due to its versatility, and it has long been the most popular breed of cattle used for work in New England. The cows produced ample amounts of milk for the farm and the steers were trained as oxen and later used for beef.

The breed is hard to beat for its disposition and willingness to work. It is preferred over the Beef Shorthorn because it is more angular and athletic and does not put on excess weight as easily, especially when young.

Milking Shorthorns are medium in size and do not grow as large as Holsteins, but generally carry

The Milking Shorthorn is prized for its even temperament, good muscling, and longevity.

more flesh and are easier to match in color. Individuals may be red, red and white, all white, or roan (alternating red and white hairs). It should be noted, however, that many breeders of Milking Shorthorns in the United States have been crossbreeding with Red Holsteins to improve milk production in the breed. This has resulted in animals that are larger, as well as being less heavily muscled with finer bones, particularly in the legs.

Milking Shorthorns have historically been called Durham cattle, and many ox teamsters in New England still refer to them by this name. They are popular among young teamsters, as they can still be purchased as calves from dairy farms. In other regions Milking Shorthorns are nearly impossible to find.

Texas Longhorns come in a variety of colors and patterns, but their lean bodies and long horns make them easy to recognize.

TEXAS LONGHORN

Origin	United States
Color	any color or combination
Mature Weight	1,800–2,000 lbs. (815–900 kg.) (average)
Height	medium
Temperament	active and alert
Tractability	10
Other	impressive horns

The Texas Longhorn originated from Criollo cattle brought as early as 1609 by the Spanish, via Mexico, into the territory that is now Texas. Criollo cattle are the same as the Corriente breed used today primarily for roping and steer wrestling in rodeos. Small and nondescript, they were some of the first cattle brought to this continent by the Spanish. The Criollo evolved through decades of natural selection in the harsh environment of the wide-open and arid range.

While early examples of the breed were rangy and varied greatly in size and quality, modern Texas Longhorns have become a true beef breed. They differ from other modern beef cattle in being leaner and having no color standard. Of course, they also grow a tremendous set of horns.

The Texas Longhorn is one of the more common breeds used for oxen outside New England. It is popular for parades, historic reenactments, and movies, especially in the western United States. In New England, Longhorns are seen primarily in pulling competitions. They are medium sized and easy keepers. Their temperament is sometimes considered challenging, probably the result of running free with their dams or of training that would be considered inadequate by New England standards. It should be noted that their large horns can be difficult to manage in the traditionally sized neck yoke or the head yoke.

OTHER BREEDS

Many other purebred cattle are used as oxen in New England, including the Simmental, Gelbvieh, Salers, Scotch Highland, Piedmontese, Pinzgauer, and Guernsey. Any one of these breeds can make fine oxen. Some breeds may be hard to find in the northeastern United States, a predominantly dairy area. Others, like the Guernsey, are sometimes considered too lightly muscled, lack the ruggedness desired in a draft animal. The Scotch Highland, with its long hair, cannot tolerate working in warmer weather.

Popular Breeds around the World

A few breeds are worth noting because of their popularity in nations around the world. Globally the most popular cattle used for draft purposes are probably the Zebu, known in the United States as the Brahma type.

The word *Zebu* refers to a species of cattle, rather than a breed. Just as hundreds of European breeds belong to the *Bos taurus* species, hundreds of breeds belong to the Zebu or *Bos indicus* species. The characteristics of the Zebu include a large hump (composed of fatty tissue that becomes larger when the animal is well fed), greater tolerance to heat and humidity, and the ability to sweat more efficiently than *Bos taurus*. Finally, the Zebu cattle also have the ability to develop some resistance to internal and external parasites.

In Asian countries, Zebu breeds are numerous, with many draft characteristics that have been maintained for centuries. In India, for example, breeds like the Mewati are known for their powerful frame and their docility as workers. The Nagori is a famous trotting breed, prized for road work. The Bargur is a fiery, restless, and difficult-to-train draft breed. The Sahiwal, common in Asia, Africa, and Australia, is noted for its milk production and fine beef qualities, making it popular to cross with local breeds.

In Africa, many breeds are generally referred to as Zebu animals. Imported breeds influenced the Madagascar Zebu, a good-sized triple-purpose breed. The preferred breed for use on the farm is the small East African Short Horned Zebu, with its small and compact but rugged body.

Other breeds worth noting are the Ankole and Watusi from East Africa, with their giant horns. They are worked by some farmers, but are considered inferior by others because they have no humps, an opinion resulting more from yoking technique than working ability. However, the Ankole and Watusi are smaller-boned than the Zebu and have less muscling in their rump and legs than Zebu of similar size.

West Africa has the N'Dama, one of the humpless Longhorns. It has a great tolerance to trypanomiasis, an important attribute in areas that are infested with the disease spread by the tsetse fly (and therefore not a problem — or advantage — in North America). The N'Dama is a solid yellow, fawn, dun, or khaki color with long horns that turn upward. It is a tiny animal: adults may weigh only 400 to 600 pounds. Yet this animal is rugged, active, and agile, and anything but docile when range-reared. N'Dama are usually worked in teams as bulls with simple head yokes.

From western to central Africa another common breed is the Fulani. Noted as a triple-purpose breed, it grows larger than the N'Dama and has better milk production. The Fulani weighs up to 1,000 pounds and is often used for crossbreeding. It is a large breed by African standards, and has the additional benefit of being more trypano-tolerant than other Zebu-type cattle.

Common Crossbreeds

Many teams are not made up of purebred animals. Some teamsters who are eager to combine the qualities of two or more breeds seek crossbred animals. Crossbreeds have an advantage for the teamster who desires characteristics of animals not found in the local region.

One typical characteristic is the ability to meet certain weight requirements of pulling classes. Many crossbreds used for oxen in New England are Holstein crosses, because Holsteins are the most common cows in the area. With the prevalence of artificial breeding, some dairy farmers are persuaded to breed their cows to non-Holstein bulls.

SHORTHORN X HOLSTEIN

Shorthorns are commonly crossed with Holsteins in New England, where Milking Shorthorns are becoming hard to find. Teamsters persuade local dairy farmers to breed their Holstein heifers to a Milking Shorthorn bull. The teamsters purchase the resulting calves at birth and bottle-feed them. The cross combines the faster-growing characteristics of the Holstein with the more flashy and athletic appearance and heavier muscling of the Milking Shorthorn. The resulting animals are often easier to color-match than pure Holsteins and may be all black, black and white, or blue roan. The crossbreeds are usually more active than pure Holsteins.

DEVON X HOLSTEIN

Crossing the Holstein with the American Milking Devon combines the larger size and more docile temperament of the Holstein with the more active nature and aggressiveness of the Devon. The result is a black animal that is slower growing than the Holstein with the more athletic and muscular appearance of the Devon. Compared to Holsteins, this crossbreed develops more muscling on the same amount of feed. They are often used successfully in lightweight pulling classes, and are popular among young teamsters not yet ready to train the more challenging purebred Milking Devon.

CHIANINA X HOLSTEIN

The Chianina-Holstein cross is popular in New England. It is usually dark brown to almost black and combines the height, growth, and activity of the Chianina with the deeper body, more solid appearance, and more docile temperament of the Holstein. It is a desirable cross among ox teamsters striving for really large but more easily controlled oxen to be used in pulling competitions.

JERSEY X HOLSTEIN

Dairy farmers frequently cross Jersey bulls on Holstein cows to create a cow that makes more milk than a Jersey, with higher components of butterfat and protein than the Holstein. The cross reduces calving difficulties for Holstein heifers yet, with more muscling and frame, the calves are faster-growing and larger than a pure Jersey. Their temperament falls between the high-strung and head-strong Jersey and the moderate Holstein. The solid black or blackish brown offspring are easy to match in color. The breed grows to be about 1,600–2,000 pounds (larger than the Jersey) and makes fine medium-sized oxen for show or farm work.

CHAPTER ROUND-UP

In selecting a team of oxen, consider how the animals will be used and the environment in which they will live. Then find a breed that best suits your needs. When evaluating animals for purchase, keep in mind the characteristics of an ideal ox. Your team will be more valuable and trouble-free if both individuals have a desirable conformation and a similar disposition.

Training older animals is more difficult than training calves, but buying older calves or yearlings eliminates some of the guesswork in trying to match two animals as a team. While no team is perfect, if you follow the guidelines outlined in this chapter, chances are you will end up with two well-matched animals that work well as a team. ★

OX BREEDS

AYRSHIRE. Known for being more high-strung than other dairy breeds, Ayrshires tend to carry more flesh than Jerseys or Guernseys do. They have large sweeping horns and are medium in size.

BROWN SWISS. The most rugged of the dairy breeds, with exceptionally strong feet and legs, Brown Swiss can be quite large, calm, and often slow to a fault as they get older. Although a good choice for a beginner, they often outgrow young teamsters.

Ayrshire

Brown Swiss

Brown Swiss

Charolais

CHAROLAIS. This all-white breed was originally bred and used as a draft animal in its native France, where it was worked in a head yoke. Oxen of this breed are less often seen in the United States, but with good training their rugged frame and heavy muscling are desirable characteristics.

CHIANINA. This is said to be the largest breed of cattle in the world. It is not by coincidence that the all-white Italian variety has gained popularity in the northeastern United States, where this breed usually dominates ox competitions. These are big-boned, tall, and heavily muscled cattle that were bred to work.

Chianina

Chianina-Holstein

Chianina-Holstein

CHIANINA-HOLSTEIN. As the huge Chianina has come to dominate many of the pulling classes in the northeastern United States, some teamsters seek a cross between it and the Holstein to create an ox with a more moderate temperament. The Chianina bull is often bred artificially to Holstein cows to create an animal for the "Free for All" or unlimited weight class in ox pulling competitions.

AMERICAN MILKING DEVON. The triple-purpose "old style" Devon is the most popular Devon breed for use as oxen. Lively, small to medium in size, they are most like the cattle originally bred for draft purposes.

American Milking Devon

Devon-Holstein

Dexter

Dutch Belt

DEVON-HOLSTEIN. Most often a cross between the Holstein cow and an American Milking Devon bull, this all-black or mostly black animal is larger than the Devon and more moderate in temperament.

DEXTER. One of the smallest breeds of cattle in the United States, the Dexter was a small farm cow, developed in Ireland. The breed carries more flesh than more modern dairy breeds, and can make a fine team of oxen. They do require a little more experience in training and handling as oxen, as they are more independent and more willing to test you than common dairy breeds.

DUTCH BELT. More refined than the Holstein, this medium-sized breed is easy to match and frequently sought out by ox teamsters who want striking show cattle.

Horned Hereford

Horned Hereford

Hereford-Holstein

Hereford-Holstein

Holstein

HORNED HEREFORD. Common historically as a draft animal, the Horned Hereford has remained a breed of choice for many ox teamsters using a head yoke, especially those in Nova Scotia, Canada.

HEREFORD-HOLSTEIN. This breed is easy to match in color, as all of the animals are black with a white head. Most often the offspring of Holstein cows and Hereford bulls, the breed is known for its calm disposition and medium size.

HOLSTEIN. The most common dairy breed in the United States and Western Europe, Holsteins grow to be large oxen but may not fill out their frames until they get some age on them. As the most productive dairy breed, the cows have been bred exclusively for milk production, and many of the traits that are desired in oxen, such as muscling and rugged feet and legs, have been lost. However, their temperament is usually easy to get along with, and they make fine oxen for general work and showing.

Holstein-Shorthorn

HOLSTEIN-SHORTHORN. Often all black or blue roan, this breed combines the fleshy appearance of the Milking Shorthorn with the dark color of the Holstein.

HIGHLAND. A long-haired, striking breed with long sweeping horns, Highland oxen are most often used for showing, as they are less tolerant of heat when worked hard. The team pictured here was trained to be driven with bridles and bits, which, it should be noted, is more of a novelty than common practice.

Highland

JERSEY. This small dairy breed can be easy to find and cheap to acquire in dairy areas, because the bull calves have little value. Jerseys can be high-strung, but with good training they can make fine oxen for a farm or show.

JERSEY-HOLSTEIN. Breeding a Jersey bull to a Holstein cow can minimize calving difficulties, especially with heifers. The resulting crossbred cow's milk has more protein and butterfat, and the bull calves (often cheaper than Holstein calves) make fine oxen. The Holstein genetics moderate the Jersey's temperament and add frame and size. The all-black or blackish brown animals are easy to match, often seen as both show and farm teams.

MILKING SHORTHORN. For nearly a century the breed of choice for teamsters in the Northeast, these are medium-sized, carry more flesh than other dairy breeds, and have a desirable temperament as work animals.

Jersey

Jersey-Holstein

Milking Shorthorn

Randall Lineback

RANDALL LINEBACK. This is a rare breed of cattle in the United States. Small in size, it is known for its unique color markings: a blue roan color, a distinct white line on the back, and a brockle face. They are most often sought for use in show, and their temperament is usually easy to work with.

TEXAS LONGHORN. Developed in the United States from Spanish cattle, this beef breed comes in many colors and has the longest horns of any cattle bred here. The animals tend to be harder to manage than other breeds, but they are long-lived and have made their way into the smaller weight classes in pulling contests in the northeastern United States.

ZEBU. This is a common name for *Bos indicus* cattle, largely originating in Asia and particularly common in India. In North America, the Brahma was a *Bos indicus* breed developed for beef production. The Zebu are actually a different species than the European breeds, which are *Bos taurus* cattle. Zebu cattle come in hundreds of different breeds and are more commonly used as oxen around the world than are *Bos taurus* cattle.

Texas Longhorn

Zebu

Ross and Luke Conroy

Berwick, Maine

My sons Luke (age six) and Ross (age twelve) have always been interested in our farm and its animals. For as long as Luke can remember, I have had only American Milking Devon working steers or oxen. The Devons we have kept, however, were not animals they were allowed to play with; my sons could not even enter their pasture without an adult. The cows were very protective of their calves, and the animals would have taken advantage of the boys if they had a chance.

When the Devon steers were yoked up, both boys learned how to drive them under my supervision. They often went for cart or sled rides, and they helped me haul our firewood out of the forest. In the yoke, even the best-trained Devons were quick to take advantage of the boys and just walk or run off, as they would with any teamster who did not know what he was doing.

Ross remembers a pair of Dutch Belted steers we had raised on a bottle from babies, and he says he enjoyed that team more than any of the Devon steers. He showed this team at a fair. I was quietly hoping he might take to training and driving his own team. However, when the opportunity arose, I purchased more Devon steers, and over the next four to five years the interest he had in steers seemed to disappear with the sale of the Dutch Belted team.

The Devons are fast-moving cattle, and some teamsters have told me that they are not an old man's team of choice for this reason. They are also not the likely choice for a budding young teamster, for all of the reasons above.

I decided that the breed I began with, 30 years ago, might be the best way to interest Luke in train-ing oxen, since I did not believe Ross had any interest. I purchased three Brown Swiss calves, and both boys quickly fell in love with them. The boys enjoyed bottle-feeding the calves, and they halter-trained all three calves while they were still on the bottle. Every time I said we ought to work with the calves, the boys jumped at the opportunity. With the boys always interacting with them, the calves almost seemed to train themselves. They were calm, slow moving, and fun for the boys to work with. I never had to worry about them running off or taking advantage of the kids.

According to Ross, "The Brown Swiss are easier to work with, as they are not as jumpy, and just more friendly than the Devon calves." He went on to say, "It seems like the Brown Swiss understand you more than the Devons."

I have to admit the Devon calves we've had have always run with their mothers on pasture until they were weaned. This always makes for more challenging training, no matter what breed is chosen. Even with the same training, however, Ross summed it up well when he said, "The Devons are more likely to want to do their own thing."

According to Luke, who wrote his first story in kindergarten about Bill, his favorite of our three calves, "Brown Swiss are a lot easier to train, and that is basically it."

2

HOUSING
YOUR
OXEN

ousing for working cattle varies as much as the cattle do themselves. Oxen do not require elaborate housing. Cattle in a variety of climates are kept outdoors most of the year. Pastures may provide both an excellent environment and a readily available feed source. Keeping a team on pasture also minimizes the need for providing bedding, handling manure, and hauling feed. On the other hand, this method does have its drawbacks.

While pastured oxen are easy to care for, they may not get the regular handling essential to maintaining a well-trained team without some effort. Good pasture may provide too much feed for mature oxen that are allowed to graze all the time. Cattle that spend four to six hours working each day may not have this problem, but mature oxen that work less often will soon become obese. Regularly moving them to a corral or barn may be necessary to prevent overfeeding.

Finally, housing for cattle should have some flexibility, especially if you purchase your animals as calves and grow them to maturity, or if you keep different breeds or species in the same facility.

Housing

An ox team in New England was traditionally housed in a barn during the winter. The cattle were worked on most days and spent nights in the barn, near other cattle for ease of feeding and cleaning. During bad weather many teams may have stood idle for days or even weeks.

Neck chains, with the cattle tied to wooden upright poles, were typical in the late nineteenth century. In earlier periods crude stanchions or wooden collars were used to hold the animals in the barn. Few barns had loose housing.

Oxen in New England are housed in much the same way today. They are often kept in barns where they are tied individually. Neck chains or leather collars with short chains and swivels are preferred because they are flexible and comfortable. Stanchions work, but they restrain an animal's movement, especially when lying down. A young team of oxen quickly outgrows stanchions designed for cows.

An ox is usually tied near his teammate in the same position as he works in the yoke. This practice helps familiarize the animals with each other. Tying also aids in teaching the animals how to step in and step out during grooming and barn cleaning.

SPACE REQUIREMENTS

Listing rigid dimensions or recommending specific housing designs would be futile, since the sizes of oxen and the budgets of teamsters vary widely. If your cattle will be housed indoors, make your barn or shed large enough for your animals. Many beginning teamsters starting with young animals do not realize how big the animals will get.

An old converted dairy barn may work fine for young steers, but the animals will soon be standing in the gutter and be too tall for the stanchions. An ox grows to almost twice the size of a cow of the same breed and therefore requires more space.

Cattle are tied in the barn as they would stand in the yoke.

Design doors, stall dimensions, and wood floors with the largest oxen you might keep in mind.

Since most oxen have horns, animals housed in cramped quarters should be tied. If you do use loose housing, be sure to provide ample space for all animals. Minimize corners and tight quarters to allow the less dominant animals to escape if chased by larger ones.

Tie stalls come in many possible designs. Simple stalls or tie-ups that are similar to but larger than those used in dairy barns work best. Dairy stalls are about 4 feet wide and 6 feet long, not including the 2 feet of manger space, called "lunge space," which allows cattle to lie down in a normal position, and to throw their heads forward as they get up. Stall space for medium-sized oxen should be about 4 feet by 8 feet. Stalls may be smaller for calves and the smaller breeds like the Dexter, and must be a lot larger for the huge breeds, such as the Chianina.

VENTILATION

Modern dairy farmers have moved away from the stanchion barns with a few small windows far above the animals' heads. Dairy cows produce more milk in an environment that is comfortable and cool. Oxen benefit from the same treatment. The animals get the greatest benefit from air movement when windows are arranged on the wall they face while tied or in their stalls, allowing air to flow past the animals in hot weather.

Cold weather and snow do not affect cattle as much as wind and cold rain, which causes them to run for shelter. Even when cattle are tied in a barn or stall, a dry environment and excellent ventilation are critically important for good health. Cattle are more comfortable in an open shed that is draft-free and dry than in a stuffy, hot, and humid barn where the animals are crowded and damp. Leaving barn doors and windows open during winter is healthier than shutting the barn up tight.

FLOORING

Keeping oxen indoors entails frequent cleaning of stalls. Cattle prefer to sleep on dry bedding. They do not like to lie on wet, frozen, or muddy ground.

Steers, oxen, and bulls "piddle in the middle," or urinate beneath themselves. Gravel, dirt, or sand floors offer the best drainage and animal comfort, but they are hard to clean. Wood floors are more comfortable than concrete but require lots of bedding, and when the wood rots the floor must be replaced.

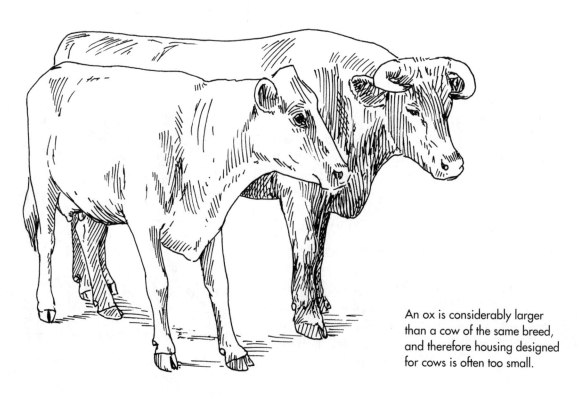

An ox is considerably larger than a cow of the same breed, and therefore housing designed for cows is often too small.

Concrete is easy to clean but offers no drainage and little animal comfort. One way to overcome this problem is to put rubber mats over the concrete and put a small drain, or drill holes in the concrete, to drain away urine.

Concrete and other hard surfaces are detrimental and can result in lameness in animals that are prone to swollen hocks and diseases of the hoof. Concrete floors require more bedding than other systems. Extra bedding helps prevent injury to the hocks and legs and prevents urine from making direct contact with the skin.

A barn with gravel under the oxen and concrete or wood in other areas offers the best combination of surfaces. The concrete alleys and floors are easy to clean, but the steers are standing on a natural surface, which is best for their comfort and well-being. Heavy older oxen benefit the most from a floor that offers a more natural surface.

STALLS AND RESTRAINTS

Neck chains or leather collars, attached by means of a short chain to a steel ring around a rugged vertical steel pipe or hardwood pole, offer maximum

Individual feeding is important, especially with calves.

Make sure the neck chain, collar, and pipe are strong enough to hold an animal that fights it.

Older oxen often need to be limit-fed, so a comfortable tie-up can be part of their routine.

Outdoor Management

No single ox-housing design works best for all teamsters. Keeping animals tied in a barn requires daily handling of the animals, giving them little freedom but critical and frequent human contact. However, regular handling may also be incorporated into outdoor systems of management. A corral may be used to keep animals in close contact with humans. Capturing and tying the animals at feeding time is another way to incorporate regular handling into your management program.

Some of the healthiest and cleanest oxen spend most of their time outdoors. Animals with freedom to move around in fresh air and sunshine are less likely to have ringworm, respiratory problems, and external parasites such as lice and mites. These health problems occur more frequently in barns where cattle are kept in tight quarters.

comfort to a restrained animal. I do not recommend steel pipes or bars that run horizontally in front of the animals' necks. While a horizontal pipe a few feet off the ground may offer good restraint and keep the animals from stepping on their feed, it creates calluses and sores on the animals' necks, which will interfere with or be irritated by the yoke.

Small calves in big stalls or tie-ups will turn around to see what is going on, often resulting in manure and urine in the feeding area. Calf hutches work well for bottle-fed calves in all seasons of the year. If the animals are kept in the barn, calves must be separated to minimize the transmission of disease.

If larger animals try to turn in their stalls, temporary dividers may be necessary to restrict movement. Dividers suspended from the ceiling are more comfortable for large animals and are less likely to be destroyed than temporary dividers on the ground.

To keep feed clean and dry, design the stall with a slightly raised feeding area. Although high feeders will keep the feed clean, the feeding area should not be similar to that in a horse stall. This is because cattle do not like to reach up or hold their heads high in order to eat. Cattle prefer to feed with their heads down. The feeding area should be about 6 inches higher than the surrounding floor.

These tie stalls have gravel or a platform that raises steers 2" to 4" above floor level, with a manger 3" to 4" above the platform. Steel pipes slide into holes drilled into two horizontal wooden beams and can easily be pulled out for replacement if bent by a strong ox. At feeding time each animal's collar is snapped to a short chain attached to a steel ring around one of the poles.

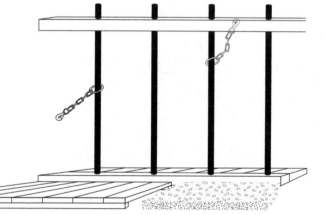

Janet Conroy
Berwick, Maine

Is a barn or a shed a better housing solution for oxen? Many teamsters, especially in New England, believe in tying up oxen in a barn with their teammate daily. While regularly tying animals together in a barn is good for training and makes individual feeding easy, in some ways it is tougher on the animals and people who care for them.

With this in mind, I decided to get my wife Janet's perspective on housing oxen. She has been a partner in most of my endeavors with oxen, if (she would admit) not always by choice. When we bought our little farm in Berwick, Maine, we started with a shed for our pair of Dutch Belted steers because that was all we could afford. We had our hay under plastic and hauled water from the house all winter. It was less than ideal for feeding and watering. However, the animals stayed cleaner, we used less bedding, and we did not constantly have to shovel their manure and wet bedding.

The next year we designed a barn for a pair or two of oxen, as well as some pens for future animals. Janet said I could plan the inside, but she wanted a say on the outside. She wanted the barn to match our house in trim color, siding, and roofing. She also wanted to minimize the number of trees I cut down, so the barn would fit nicely into our wooded, riverside farm.

Our final barn was 24' by 24' with a "salt-box" roof and a hayloft that might hold 300 to 400 bales of hay. I knew the barn size and hayloft would limit the number of steers I could keep.

We kept the sheds we had originally built for the Dutch Belted steers; in fact, 17 years later they are still there, having been used every winter to house oxen. When I travel, often for days, weeks, or even months, Janet has found a number of advantages in the sheds.

"If you [Drew] are gone, I have to clean out the shed only once every few days. The steers stay cleaner, as they are not lying in their own urine-soaked bedding. We use less bedding, which can be expensive, and they have fresh air, especially in winter. I think they prefer to be out, rather than tied up all the time. The easiest time is when they are on pasture in the summer, because then there is no shoveling of manure."

Janet also notes some disadvantages: "We have to haul water and hay to the paddock and sheds from the barn. I hate doing that, especially in winter. An automatic waterer would help, and a small hayloft in the sheds. Finally, the hitching rail near the sheds where you individ-

ually feed them is not something I like either. It's not fun trying to feed grain and tie the Devon steers up with those horns."

We have always had tie-ups in the barn for at least two teams. Janet's comment: "There is more cleaning out, and we use more bedding. The way it is set up you have to get up near their heads to feed them in a crowded space. The advantages are that the water is nearby and the hay is overhead. Even so, I would like warm water in the barn, more lights from the house to the barn (it is a few hundred feet from our house), and a staircase rather than a ladder to the loft."

Janet knows the barn is more accessible for taming calves, especially for our boys. As with any housing system, however, there are advantages and disadvantages. I learned a few things asking my wife after 20 years what she thought of our ox housing.

Fences

Whether oxen are kept outdoors or only occasionally let out, they are usually fenced in. The fence should be of a type suitable for beef or dairy heifers and steers. Heed the old adage that the grass is always greener on the other side of the fence: No matter what type of fence you put up, your animals are more likely to remain in their pasture if it offers adequate feed, compared to a pasture with better grazing on the other side of the fence.

Solid wood or woven wire makes the most rugged and trouble-free fence, provided it is tall enough that the animals cannot graze over it, and strong enough that the animals cannot push it over. A single electric wire inside the fence will keep the animals from leaning on or over it. The combination of the physical barrier provided by the wooden fence and the psychological barrier provided by the electric wire is probably the most foolproof system, but it requires a substantial investment in time and materials.

Barbed wire works too, with a lower investment, but I have seen many cattle walk through even the most tightly stretched 4- or 5-strand barbed-wire fence. Cattle are much tougher than horses and more careful and deliberate in pushing through a barbed-wire fence. I once had a pair of Brown Swiss that would push against barbed wire until a couple of strands broke, then amble out of the pasture. The act was intentional and they never received more than a few small cuts. Apparently the greener grass on the other side was worth it. I have chased more animals that escaped from barbed wire than from any other fencing system.

On my farm I use electric fence almost exclusively. My cattle remain outdoors for most of the year. A high-tensile multistrand fence, with a powerful charger and good grounding, is hard to beat for animals that have been trained to respect it. I have never had animals get out, unless the fence charger was struck by lightning or I turned it off and forgot to turn it back on. In both cases it was days before the animals learned that the fence was off. Even then only one animal out of the herd got out. The bawling of the remaining animals informed me of the escape. A high-tensile fence is more expensive than other fences, as the wire is heavier gauge than standard 18- to 20-gauge electric fence wire, but it also provides a stronger barrier if an animal bumps into or is chased into it.

Common one- or two-strand electric fences can be used successfully with animals that were trained to respect a high-tensile fence charged with a high-voltage, low-impedance system. In this case, one or two strands, on wooden or fiberglass posts, usually works. Sometimes, however, the animals push each other through the fence, chase a calf through, or knock over posts by rubbing on them with their horns.

Any electric fence has a number of disadvantages. When it fails, cattle are much quicker than other farm animals, such as horses, to figure out that it is off. Oxen with horns tend to touch the wire with their horns, and on dry or frozen ground they do not get a good shock through the horns. If the weed load on a long stretch of fence becomes heavy and wet, even the best fence charger fails to deliver a strong shock, resulting in animals escaping. On the other hand, a 5-foot-tall high-tensile fence with strong and long-lasting posts makes a permanent, attractive, and low-maintenance barrier fence.

CHAPTER ROUND-UP

Dry is good and wet is bad when it comes to cattle housing and ventilation. Remember this simple rule to minimize the possibility of disease and maximize your animals' comfort.

Whatever kind of housing you choose, make sure it provides a dry, bedded, and well-ventilated place where your animals can get out of wind and rain in winter and enjoy shade during summer. Equally important, the design should be one that encourages you to handle your animals regularly. ★

3
FEEDING
YOUR
OXEN

Feeding the working steer or ox is not as complicated as feeding the beef steer or lactating dairy cow. Due to the nature of diets that cattle receive in intensive management systems, most high-producing dairy cattle and feedlot steers walk a fine line between maximum production and having an unhealthy rumen, which predisposes them to numerous digestive disorders. Feeding a steer used for work differs in many ways and is easier than feeding a steer destined for the dinner plate.

The Bovine Digestive Tract

Oxen are ruminants, which means that they have four stomach compartments: the reticulum, the rumen, the omasum, and the abomasum. The rumen and reticulum share a primary digestion function, differing mostly in size and appearance.

The **rumen** is large. In a mature ox weighing more than a ton, the rumen can have a capacity of 60 gallons or more. The rumen takes up a considerable amount of space in the animal's gut and is important to the animal's overall health.

The **reticulum** is much smaller and is located in the front portion of the rumen. It is frequently called the "honeycomb," because its interior looks like a honeycomb. Occasionally it is called the "hardware stomach," because it is the part of the stomach that is most likely to be punctured by nails or wire the animal inadvertently eats.

Both compartments contain large populations of live microbes, including bacteria, fungi, and protozoa that are essential to the survival of the ruminant animal. These microbes are the primary digesters of roughage; without them a mature animal becomes ill.

The **omasum** is about the same size and shape as a basketball. It is filled with many flat layers of tissue that remove water from the partially digested material leaving the rumen.

The **abomasum,** the fourth stomach compartment, is located at the bottom of the abdomen. It functions as the true stomach, complete with digestive enzymes and hydrochloric acid.

Adult ruminants have the unique ability to consume forage and convert it to usable nutrients for body maintenance, growth, or work. A calf is born without a functional rumen and therefore cannot digest fibrous feed right away. Although the rumen

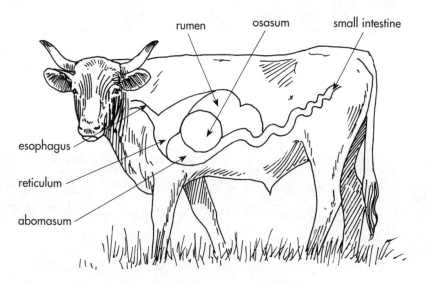

rumen   osasum   small intestine

esophagus

reticulum

abomasum

The bovine digestive tract.

develops rapidly during the first few months, and within one year is fully functional, the calf less than one year old has requirements that exceed the rumen's ability to digest feeds. The young animal thus needs high-quality, highly digestible feeds.

Feeding Young Animals

A calf's protein and energy needs are greatest during his first year of life. The term *energy*, as commonly used in discussing livestock nutrition, is defined as: feeds that fuel normal body function, metabolism, and muscle activity.

For cattle, carbohydrates are the most common source of energy. Most grains and fibrous feeds or forages are high in carbohydrates. Fats and proteins can also provide energy, but are generally more expensive sources. Although fats are energy-dense, cattle have a hard time digesting a high-fat diet. Proteins are best utilized for purposes other than energy metabolism. Using proteins for an ox's energy requirement not only wastes money but also leads to excess protein and nitrogen in the manure and urine. The result is often a strong ammonia odor in the barn.

Working steers between birth and one year require a diet similar to that fed to dairy or beef heifers destined for breeding. This diet includes allowing the calf to nurse a cow or bottle-feeding, and then placing the animal on a highly palatable and digestible ration until his rumen becomes functional.

After his first few months of life the animal must gradually make the transition to a higher level of forage and a lower level of more digestible feeds, such as milk, milk replacer, or grains. The first year is the only stage of development when a steer destined for work may be allowed free-choice forage feeding. Steers beyond one year of age, if allowed free-choice forage of better than average quality, become overweight, have a hard time working in the yoke, and will have a shortened life span.

A number of books and pamphlets are superb guides to feeding rapidly growing young animals. The texts *Beef Cattle Science* and *Dairy Cattle Science* by M.E. Ensminger (see Bibliography) are both excellent references that provide all the information you

Roughage

Oxen were historically the draft animals preferred by poor farmers and those trekking long distances. The mature steer's ability to graze and consume large amounts of coarse feeds in a short period of time, and then process those feeds throughout the day while chewing his cud, was his greatest advantage over the horse on long treks. This wonderful attribute still makes the ox popular among the poorest farmers around the world.

Steers older than one year can thrive primarily on roughage. They do not need to be fed for maximum growth or intake. Most experienced cattle owners in the United States, however, have a hard time restricting the diet of working steers. Like the American diet in general, the diet of many an ox in the United States too often leads to a shortened working life and health problems related to obesity.

Breeders have genetically selected cattle to grow quickly and fatten easily. Most cattle owners are familiar only with bovine diets designed for high production. The most common result is a working steer that becomes more like a pasture potato than an athlete. Many young steers are too fat to do much of anything but lie in the shade, chew their cuds, and look big.

need to balance the diets of your cattle at every stage of their lives. To give your animals a good start, follow the guidelines set forth in any reputable guide to feeding growing cattle and provide the best possible feed in terms of nutrients per pound and high palatability.

FEEDING CALVES

One of the most important rules in starting young calves is to make sure they are fed properly. Shortly after a calf is born it *must* receive colostrum — the first milk from a cow after she has given birth. Colostrum is more nutritious than normal whole milk or replacer and, when fed within the calf's first few hours of life, will give the newborn animal some immunity to disease. Usually 2 quarts, fed twice or three times daily for the first few days, is a sufficient amount.

Bull calves are not a dairy farm's top priority and thus may not receive colostrum. Many calves sent to market have not received adequate colostrum. For this reason, try to buy calves directly from a reputable farmer who feeds colostrum to all newborn calves. If possible, let the farmer start the calves for a few days or even weeks before you buy them. In the long run, paying a little more for well-started calves is worth the money. Starting a team with day-old calves purchased at a local sale barn at a bargain price can result in dead calves and a disappointed teamster.

After they have had colostrum for a few days, give the calves whole milk or milk replacer made from real dairy products. Cheaper milk replacers, made from plant proteins as the primary ingredient, are less digestible and may cause scours, or diarrhea.

A common cause of scours is feeding a calf too much milk at a feeding. A calf should get no more than 10 percent of his body weight in milk per day. Divide this amount into at least two equal feedings, preferably three or four, as the calf nursing a cow might feed 10 times a day. Feeding more often and offering less per feeding is easier on the calf, but make sure he gets no more than a total of 10 percent of his body weight each day. As a general rule, a hungry calf is a healthy calf. Feeding a calf until he is full is not a sound practice.

Nursing

If you have a willing nurse cow that is a high-producing dairy animal, such as a Holstein or Jersey, two hungry calves won't be able to keep up with her milk production. The result may be calves with scours due to digestive upsets from too much milk, and a cow with mastitis because the calves can't remove enough milk from her udder daily.

Another way to start young calves is to place them with a nurse cow, where they can drink at will if the cow lets them. A cow that refuses may be restrained while the calves nurse, but this practice limits the calves' access to the cow.

In a beef operation, on the other hand, calves raised by their dam usually get a good start. Training them, however, is more challenging because as mentioned earlier, they lack the human contact that comes from bottle-feeding. Beef calves often nurse the cow for six months or more. At the end of six or seven months the calves might weigh 500 or 600 pounds. Compared to bottle-fed calves, they are less likely to get scours and usually look healthier, but for a novice teamster will be a handful to train.

Whether bottle-fed or nursed, calves should have water available at all times. In hot weather, if the milk is not enough they will learn to drink water to satisfy their thirst.

Bottle feeding ensures frequent handling and important human contact, making this animal's future training easier than if it were allowed to nurse the cow.

Feeding the Growing Steer

Beyond five or six months cattle can get along on forage alone. They are efficient animals and will survive on feeds most other animals couldn't or wouldn't consume. However, if they are fed poor-quality forage they will develop the classic hay belly, or pot belly, as a way to increase their ability to consume and process low-quality feed. Although such animals may be healthy, they tend to have less energy and be leaner with prominent ribs and hips.

Fast-growing flashy animals that appear healthy during their first year or two require top-quality for-ages such as alfalfa, corn silage, or well-managed pasture. If top-quality forages are not available, you can get the same results during those first few years by feeding more grain. Young cattle of the dairy breeds usually need grain and pasture to maintain a high degree of muscling and an athletic appearance.

The American Milking Devon steers in these two photographs are three months apart in age. The younger steer was kept on milk for 14 weeks and fed a greater amount of commercially prepared calf starter. The larger steer was fed more hay and little grain, until the two were similar in size. At one year of age, as shown in the lower photo, the calves were nearly impossible to tell apart. The animal on the nigh (left) side in the yoke is the younger calf.

Monitoring Growth

The most common difficulty faced by new cattle owners is determining how their animals are growing and thriving. You can ascertain the growth rate by weighing your animals or by using weight tapes, available in most feed stores or animal supply catalogs. Calves should gain a pound or two each day, depending on the quality of feed and the animals' breed and age. The rate of gain may vary from this guideline for short periods without being detrimental.

Good health, a healthy hair coat, and hooves and horns that are relatively smooth (without hardship lines) should accompany steady growth. Animals that are growing but have rough coats, severely flaky horns, hardship lines (due to laminitis) on the hooves or horns, and too little flesh may need a higher protein or energy level.

Restricting energy, protein, or fiber may cause problems such as permanent stunting or reduced performance. During their first year, calves should get the best-quality feed available. Try not to restrict their intake of water, and give them access to free-choice forage.

In the show ring you can readily tell which owners are feeding their young animals for appearance and maximum growth. 4-Hers often wonder how to make their young animals look better. If the animal looks thin the usual answer is deworming. If deworming doesn't help, increase the protein and energy content of the diet, which is easily done by feeding more of a commercially made pelleted grain or concentrate. Feeding only low-quality forage restricts the animal's growth and alters his condition, due to low protein and energy. Good-quality forage, on the other hand, can provide the animal with most or all of his nutritional requirements. Most working steers are fed primarily forage, with a concentrate only to balance the animals' protein and energy needs.

FEEDING YEARLINGS

When your animals pass their first birthday, seriously reevaluate their diets. Dairy breeds, such as the Jersey, Holstein, and Guernsey, tend to be lean and angular. Steers of these breeds may need more grain to develop the muscling desired, and that still may not happen until they are two to three years old. On forage alone they will grow and eventually fill out their frames, but growth may take five or six years.

Beef cattle fill out more easily on forage. Breeds like the Brown Swiss or Chianina require more feed at the same weight than smaller-framed breeds such as the Hereford or Shorthorn, in part because they are such fast-growing animals.

Nutritional Requirements

As any animal grows, his dietary requirements change. Suppose you put three animals of different ages on a well-managed pasture: a two-week-old calf would starve, a three-year-old steer would look terrific, and a ten-year-old ox would be obese. In designing a diet to meet the needs of your animals and your budget, take into consideration both the animals' requirements and the available feeds.

Many people feed cattle for years without knowing the exact protein and energy level of the feeds they use. Having feeds tested — or at least using book values for protein, energy, and mineral contents — can be helpful.

A steer's nutrient requirements depend on his breed, age, weight, condition, rate of gain, and average daily consumption. Compared to young animals, older animals generally need less protein, less energy, and more fiber, and they consume a lower percent of their body weight each day.

A lean or underweight animal may require more feed or a higher-quality feed for short periods of time. Many new cattle owners are surprised to learn that you can grow animals at different rates. You can feed for no gain (sometimes done to keep an animal in the same weight class all season) or you can feed for one or more pounds of gain per day. The choice is up to the person controlling the feed bucket.

To illustrate this, the table on page 272 outlines the nutrient requirements of steers of different ages and sizes. This table shows the differences in nutritional requirements based on extreme examples. Remember that "total intake per day" is just that, the total amount of feed consumed as it is fed. Dry

matter intake is often used to measure the consumption of dairy cattle rations, because dairy rations usually contain large quantities of wet feeds.

Most ox teams in the United States are fed more hay than silage, requiring fewer calculations for "as fed" values. Most working steers should gain a pound or two per day until they reach maturity, which may be as late as five or six years. Under optimal conditions they may gain more. This table also illustrates the tremendous differences in protein and energy requirements between a 300-pound calf and a mature ox. Feed your animals based on their requirements and they will always look good.

BALANCING THE DIET

Balancing a young steer's diet is like feeding a child. You have to be sure the diet is balanced to meet the needs of the fast-growing youngster. While ruminants are great survivors, their proper growth and development depend on good feeding practices.

Balancing the diet of a steer to meet his exact requirements involves some mathematical calculations. The calculations are not complex when you're feeding a steer used for work because working animals are easy to feed compared to high-producing dairy cattle. Hard work increases only an animal's requirement for energy, which may be easily met with more grain or a high-quality forage.

While grain feeding or supplementation is important, it must complement the forage the animals consume. Too much grain and too little forage will limit a young calf's rumen development. This will also predispose them to an acidic rumen, leading to many other problems including laminitis, ulcers, and diarrhea.

Good-quality alfalfa hay with lots of leaves and small stems might have a protein content of 20 percent. Such forage is best supplemented with a low-protein, high-energy feed like corn or barley. On the other extreme, a corn silage diet will need a protein supplement like soybean meal or a prepared protein mix. A diet of primarily late-cut grass hay may need both additional protein and energy, so a corn/soybean meal mix or prepared pellet may be beneficial.

Water

Oxen need plenty of clean, fresh water. In drier developing countries, cattle often suffer from water deprivation. Cattle in cold climates may also suffer from water deprivation if their source of drinking water freezes. Oxen are sometimes restricted from normal water intake so they can weigh into certain classes for pulling competitions (much like athletes who withhold food and water to wrestle or box in certain weight categories).

Although I do not recommend restricting water intake, sometimes providing constant access to water is not feasible. Even in these cases, cattle should have access to water while working in hot weather and also during or immediately after feeding on dry forages. Some owners adapt their cattle to being watered only once a day, but cattle that are not used to restricted water intake can develop an impacted gut, similar to colic in horses.

Why Balance the Ration?

I have successfully fed many steers without taking special care to balance their diets. This comes in part from experience and from knowing what kind of growth and condition I expect. However, you can also feed for a specific growth rate, or feed the animal exactly what he requires.

To do this you need to balance the ration. Balancing the ration simply means using a combination of feeds to provide the nutrients required by the animal. The process may be as simple as making sure he gets the required protein and energy level, or as complex as balancing the diet for every nutrient including vitamins, minerals, fiber, fat, energy, and protein.

Making sure your steers get enough protein and energy is usually sufficient. For even the most basic feeding program, however, an understanding of the terms used in feed analysis and ration formulation is helpful.

Protein and Energy

Every steer requires certain amounts of protein and energy for normal growth and body maintenance. If either one of these nutrients is not fed in sufficient quantity, growth is the first thing to be affected. The animal's general health and bodily functions will be maintained as long as possible. A severely restricted diet will cause growth to stop altogether. If the animal gets too little protein or energy to maintain his normal bodily functions, his health will suffer.

Fiber

Cattle must have some dietary fiber to keep their rumen functioning normally. Without adequate fiber an animal cannot chew his cud and may therefore develop such disorders as a displaced abomasum, bloat, or ulcers.

High-fiber feeds are hays, silages, and most pasture crops. Grain hulls, or cotton lint, are also high in fiber. Grains processed with their cobs or lint are higher in fiber than grains without them. Good fiber crops include all of the grasses and legume crops, corn used for silage, cottonseed, and other whole seeds with the hulls or cob included.

Sometimes cattle used for pulling competitions are fed a restricted forage diet to help them maintain a more athletic appearance and lower gut-to-body ratio. The few grains they are fed must be as high in fiber content as possible. Whole or crimped oats provide more energy than hay or silage. Oats lack the bulk of hay or silage, yet have some fiber, providing the rumen something to digest more slowly than pelleted or ground feeds.

Vitamins

Cattle more than a few months old have a living microbe population in their gut. These microbes manufacture most of the vitamins the animals need in the rumen, so additional supplementation is normally unnecessary.

Minerals

While the mineral requirements of a working steer are not as high as those of a lactating or pregnant animal, a serious deficiency of minerals can limit growth and production. Providing a trace mineral salt block or feeding commercially prepared pelleted grain provides most of the minerals that might be lacking in locally grown feeds.

Forages

Most working oxen and steers past one year of age can derive all, or nearly all, of their nutritional requirements from good-quality forages. The only exception might be minerals, which are found in the forages but can be deficient depending on the soils in the region where the crops are grown.

HAY

Hay consists of sun-dried grasses or legumes that are baled and stacked or stored loose. Early cut hay (or hay that is leafy with few stems) is more nutritious and palatable for all animals than hay cut later in the season. Early cut hay is higher in protein and energy and is relatively free of stems, making it more desirable for young animals. It is also more likely to cost more than hay that is coarse and stemmy. Legume hay such as alfalfa, red clover, or birdsfoot trefoil is usually more costly than grass hay because of its higher protein and mineral content.

Weather damage and improper curing reduce the nutrient content of hay and often reduce its palatability, as well. Some animals refuse hay that has been weather-damaged, was cut late in the season, or is moldy.

Quality hay is green, has a pleasing aroma, is free of foreign materials, and is leafy more than it is full of stems. The more stems, the lower the nutrient content of the hay. Use the stemmy hay to feed mature oxen, and keep the leafy hay for the younger animals. While hay makes excellent feed for oxen, growing animals may need to receive supplements of grain concentrates or corn silage, due to the relatively low energy content of anything but the highest quality hay.

SILAGE

Silage is made by cutting green forage and packing it tightly in a silo. Packing silage creates an airtight and later an acidic environment in which spoilage-causing bacteria cannot survive. Ensiling forage does not improve its quality, but it does allow the forage to be preserved in such a manner as to capture a maximum amount of nutrients from the crop.

Many different crops may be ensiled. Properly made silage makes excellent feed for oxen. It is much higher in moisture content than hay. It should have a pleasing aroma and should not smell of mold or rot.

Grass silage is common in northern areas where legumes, grains, or corn cannot be grown. It allows the producer to harvest quality feed where drying is difficult or impossible. Silage retains most of the nutrients found in the original grass stand. While ensiling cannot improve the quality of the forage, it may enhance its digestibility. For young steers silage, like hay, is usually supplemented with grain concentrates in order to achieve the desired energy level.

Legume silage is higher in protein content than grass silage harvested at the same stage of maturity. Since early cut legume silage may be very high in protein and low in fiber, the feeding of grass hay can help promote a more balanced diet (based on total protein content) and yield better rumen health.

Corn silage is common in areas where dairy cattle are raised. It is usually cheaper than grass or legume silage and makes an excellent feed for steers of all ages. Corn silage in a steer's diet complements good-quality alfalfa hay because it is high in energy and the alfalfa is high in protein. Where these two feeds are readily available a feeding program that incorporates both can be economical.

With the exception of wrapped-round bales (bale silage), making silage requires more machinery than many small farmers would care to invest in. The disadvantage to purchasing silage is that it spoils quickly in warm weather and may require a special storage facility.

Other crops such as sudangrass, sorghum, millet, and cereal grains can be made into good quality silage that might be economical, depending on availability and price. These silages tend to be lower in protein content than legume silage and lower in energy than corn silage.

PASTURE

Pasture is the natural feed for all cattle, and oxen are no exception. A properly managed pasture can be economical, easily maintained, and nutritious for animals at any age. Pasture quality varies based on soil conditions and rainfall, but cattle are adaptable. Pasture provides not only good feed, but also exercise and a clean environment.

The only precaution in using pasture is to make sure you acclimate your animals to it over a few days or a week. Putting animals on a lush spring pasture after a winter on late-cut dry hay can result in severe rumen upsets, laminitis (especially in heavy steers), or even death. Bloat (caused by consuming young succulent legumes) and nitrate poisoning (the result of a severe overload of nitrogen in the blood) are also common problems in animals that are not acclimated to early-growth grass or legume pasture. Fertilizing, reseeding, and even drought conditions can change the composition of pasture and make it hazardous to animals' health, if they are not slowly acclimated to this diet change.

Pasture should be managed to provide the right type of feed for the animal. Older mature oxen can do well on a run-down, overgrazed pasture, while calves need the best pasture available. If only one pasture is available, divide it up so that the animals with the highest nutritional needs graze first and the animals with the lowest nutritional needs eat what the younger animals have left behind.

Some ox teamsters are reluctant to pasture because it may allow the animals too much freedom from human contact. If properly and frequently handled, pasture-reared oxen can be as well trained and managed as animals kept tied in a barn.

Grains or Concentrates

Hard work increases an animal's energy requirement, which can be met by adding a little grain to the diet. Dairy breeds usually need some grain in order to maintain a high degree of muscling as young working steers or oxen.

Corn is a high-energy, low-protein feed often used to supplement good-quality grass or legume hay diets that are typically low in energy. Whole corn has a higher fiber content, but much lower digestibility, than ground corn. Ground corn is more digestible, but if fed in excess is more likely to cause digestive problems. Corn, particularly processed corn, is sometimes called a "hot" feed because it readily ferments in the rumen.

Oats are fed to steers as an energy supplement. Oats are higher in protein than corn, but have less total energy. Oats are commonly fed as a major ingredient in the diets of competition animals. While whole or crimped oats do not meet *all* nutritional and fiber needs, they do satisfy most of a steer's protein and energy requirements and they stimulate the rumen.

Barley is similar in feeding value to corn but higher in protein and more easily digested. It is an excellent feed, but its easier digestibility can lead to digestive upset in a diet that is changed too rapidly.

Bakery waste (pulverized stale bread and donuts processed into a meal) and other by-product feeds make excellent supplements, but vary in nutrient content. Strive to feed primarily wastes that are free of foreign material or spoilage. Knowing the nutrient composition is essential if you want to use this type of feed, the primary advantage of which is low cost.

Soybean meal is a common supplement to diets that are low in total protein. It is often fed to boost the protein content of diets based on corn or corn silage.

When steers and oxen are allowed to free-will graze on good-quality pasture, the nutrient requirements for calves will not be met, while older animals like the white ox will become obese.

Brewer's grain, either wet or dry, is an excellent source of protein. Wet brewer's grain is hard to find in some areas and, for just a few animals, is more difficult to manage. It may be mixed into silage or chopped hay, provided it is fresh and not spoiled by heat or fermentation. Dried brewer's grain is more expensive but easier to work into a simple ration.

Beet pulp is often fed to cattle to provide both energy and protein. It may be fed either wet or mixed dry into silage. In some areas beet pulp is a cheap source of concentrate to supplement poor-quality roughage.

Wheat middlings, a component of many pelleted rations, offers both a high level of energy and some protein.

Commercial protein mix is available to supplement rations based on both forage and grain. Make sure the commercial mix complements what else you are feeding. A diet consisting of mostly good-quality alfalfa rarely needs a protein supplement, but might benefit instead from an energy supplement.

Commercially prepared pelleted feed, available in 50- or 100-pound bags or purchased in bulk, must be appropriate for your animals and should be reasonably priced. Since any grain can be pelleted, be sure the pellets you purchase contain the nutrients your animals need. Most pellets include a variety of different grains. These are mixed to meet the minimum nutrient composition specified on the feed tag. Ingredients for making a particular pellet formula may change as dictated by feed prices and availability at the feed mill. Some feeds are classified as sweet feeds because they contain molasses. Others are simply called 12-, 14-, 16-, 18- or even 32-percent crude protein pellets. Carefully consider the protein content when planning your growing steers' feeding program.

Feed Additives

A number of additives are available to supplement the diet of cattle. The three main additives used for oxen are urea, mineral mixes, and salt.

UREA

Urea is a chemically manufactured source of nitrogen, much like the urea used in a garden to supply extra nitrogen to plants to make them grow better. Fed to a ruminant, urea makes growth of the microbes in the animal's gut more efficient. With additional nitrogen the rumen bacteria and protozoa grow faster to produce more natural microbial protein.

People who keep animals other than ruminants are sometimes warned against using urea, because an animal without a rumen cannot tolerate it. Novice cattle owners are also sometimes discouraged from using it. But urea is naturally present in the rumen and is safe when fed in very small amounts. It should never be used like a protein supplement and fed directly to the animal, yet feeding your cattle commercially prepared silage or pelleted grain containing small amounts of urea is perfectly safe. Feed urea only to animals requiring additional protein and be sure to follow recommendations on the feed label.

MINERAL MIXES

Minerals may be in short supply due to the nature of the diet or to local soil conditions. Animals fed a commercially made grain mix, manufactured by a local company, will likely have minerals added that are appropriate for the region in which the feed is sold. To be sure:

★ Check the feed label.
★ Describe your feed program, specifically the forages your animals consume, to your local feed dealer. Most dealers will provide information as to what you ought to include in the diet.
★ Find out what the local beef producers use for a mineral supplement. A supplement designed for beef heifers, cows, and bulls usually works for oxen in the same region of the country.

Never feed steers or oxen free-choice minerals such as calcium and phosphorus that are designed for high-producing dairy cattle. Diets with high levels of alfalfa are already high in calcium. Animals that are fed primarily alfalfa hay, pasture, or silage should not receive additional calcium supplementation.

High levels of calcium and phosphorus may create problems for steers or oxen. The problem is compounded when the calcium-to-phosphorus ratio is improper. A ratio of 2:1 or 1:1 generally causes no problems. The main result of an improper balance of these two minerals is urinary calculi (small calcium deposits that lodge in the penis). In a severe case expensive surgery may be required, resulting in a steer that urinates out the back like a heifer. Surgery is sometimes used to save the life of a beef steer until slaughter, but makes the much-longer life of an ox quite miserable.

> High levels of calcium and phosphorus may create problems for steers or oxen.

Mineral deficiencies, except for salt, are not likely to show severe physical signs. Some micro-minerals such as selenium and fluorine are required in small amounts, but when fed in excess can create problems. Feeding minerals that are not tailored to local conditions creates serious problems. If your oxen are fed homegrown forages and grains, they'll usually do fine with a trace mineral salt block designed for beef cattle breeding herds fed a similar diet.

SALT

Salt is the most common mineral deficiency in working steers or oxen, especially those on pasture. Animals lacking this element will lick things in their environment and may chew wood or eat soil in an attempt to get the salt they need. The ultimate result is an unthrifty animal with no appetite. Supply a salt supplement in any form, loose or block, and the problem is solved.

Matching Feeds to Needs

To balance your steers' diets based on protein or energy, use both the Nutrient Requirements table and the Nutrient Composition of Feeds table (see appendix III), and follow these steps:

1. Choose the feed ingredients your animals will receive and either:
   ★ Have each feed analyzed. Most land grant universities will do this analysis or at least direct you to a lab with the capability.
   ★ Use a table to get values of feeds based on average samples previously analyzed.
2. Use the Nutrient Requirements table to determine each steer's individual requirements for protein and DMI (Dry Matter Intake). Every animal has different requirements based on his weight, age, breed, and desired growth rate.
3. Mix the feeds and present them in a way that will most appropriately meet your steer's nutritional requirements. Just because a diet meets an animal's needs on paper doesn't necessarily mean the animal will eat it. To offer a few examples:
   ★ Animals that have become used to good hay may refuse poor-quality hay unless it is presented with molasses to make it much more palatable and higher in energy.
   ★ Ground grains fed outdoors should be presented in a trough; fed on the ground they may blow away or be washed away by rain.
   ★ An animal that finds dry beet pulp unappetizing may find it more palatable when moistened.

Additional information is available in the textbooks mentioned in this book's bibliography and in literature from your local agricultural extension agent. Computer programs that calculate cattle rations are available from, among others, the National Research Council for Dairy Cattle Nutrient Requirements. If you do not use a computer program, use Pearson's Square to help you mix feeds to meet your animals' nutrient requirements. (See appendix III.)

Feeding Mature Oxen

One of the greatest challenges to feeding mature oxen in the United States is limiting their diets. Many oxen are fed too much. The mature ox that weighs 2,200 pounds needs to consume only 1.7 percent of his bodyweight per day. The 300-pound calf needs to consume 3.5 percent of his bodyweight.

A restricted diet for mature oxen means limited hours on pasture and more time spent in the barn or dry lot without feed. Estimating and controlling how much oxen on pasture consume is more difficult than watching their body condition and limiting their time on pasture.

Oxen are great work animals because they can sustain themselves on coarse roughage eaten within a short period. Historically the ox did not compete with humans for valuable grains but survived on forage and straw or grain residues. In many countries oxen are still fed in this manner and are in better health than the young obese oxen seen at shows in the United States.

Large obese animals are predisposed to lameness due to stress on the hooves. Lameness in huge oxen

Mature cattle put on weight easily, and being overweight is detrimental to their health.

Mature cattle carrying adequate flesh with the hips and some ribs visible will live longer than overweight cattle.

is one of the greatest hazards to their health. You can help prevent lameness by maintaining adequate but not excessive weight on your oxen.

FEEDING OLDER OXEN

Occasionally an ox more than 10 or 15 years old has a hard time maintaining his bodyweight on pasture or forage alone, usually due to teeth that are worn out or missing. Even when the animal's nutrient requirements are met according to recommended guidelines, the older animal may need additional supplementation.

You can often extend such an animal's productive life for a few more years by providing him with feeds that require a minimum of chewing and are easy to digest. For an animal with poor teeth the addition of grains or other easily digested feeds such as beet pulp might be necessary. If you can keep a team to the age of fifteen, you have done well.

COWS FOR WORK AND REPRODUCTION

In many developing countries cows are commonly used for draft purposes. Even though they

An ox carrying too much flesh becomes "patchy" in the tail-head, as the fat accumulates around the tail.

are smaller than steers, cows can be more efficient because they may fulfill three purposes: work, milk production, and producing young. These are wonderful assets for a small-scale farmer struggling to survive on a minimal amount of land. Feeding such animals is doable, but they require a more nutrient-dense ration than a steer of the same age or size.

Working cows that are not fed properly will quickly lose weight, and their milk production and reproduction, in that order, will suffer. The challenge is to find a way to feed them with locally grown feeds or food by-products that meet their nutritional needs.

When fed according to the guidelines for high-producing dairy cattle, working cows are effective triple-purpose animals. It goes against some cultures, however, to feed grain to cattle. Creativity is therefore required to find suitable feeds, and a basic understanding of feeding principles is essential to the process. In some circumstances applying those principles on a daily basis can be a challenging task.

Bovine Feeding Behavior

Cattle are normally aggressive eaters, and the typical steer or ox consumes feed readily and quickly. At times an animal's behavior may not be aggressive, such as when the feed is of lesser quality than the animal has been used to. If a pair of steers have been eating excellent quality hay, maybe a second or third cutting of a grass or legume, they may be reluctant to eat good or average quality hay that is more stemmy or less palatable. Such hay may be more appropriate for a mature ox but may be all you have available for the steers. Cattle will characteristically refuse good quality feed that is less desirable than feed they have been receiving, but before long will realize that they will go hungry if they do not eat what you give them.

If the feed is poor in quality, cattle can be reluctant to eat it even when given no other options. They will pick through the hay, silage, or grain to find the most palatable portions. They may waste a lot of feed and will likely lose weight. The teamster must understand the difference between lower quality feed and poor quality or spoiled feed.

Cattle may be reluctant to eat a new type of grain, silage, or hay when first presented with it. As long as the feed is palatable and not moldy they should soon acclimate to this new taste. Cattle that have never eaten silage may take a week or more to acclimate to it. They also need time to acclimate to hay that is of lesser quality than they are used to.

As long as the feed you are offering is not spoiled or moldy, do not give your cattle any options other than to eat what you have provided. If you give in to their preferences they will manipulate your conscience (like kids who prefer dessert over dinner). Cattle that are offered coarse reed canarygrass hay versus fine Kentucky bluegrass and clover hay, for example, always prefer the finer and leafier hay to the coarser hay.

. . . .

Do not make your cattle consume feed if their behavior tells you that something is wrong with it.

. . . .

WARNING SIGNS

When animals are completely reluctant to eat some type of hay or grain that they normally consume, something is wrong. The hay might be unpalatable (too many stems or the hay was rained on too many times) or the grain (even commercially prepared pelleted feed) got spoiled or moldy. Do not make your cattle consume feed if they tell you through their behavior that something is wrong with it. Hay or silage that is moldy may be toxic and can kill cattle that are forced to eat it.

If one animal in a herd or team does not eat while the others eat, usually the feed is not at fault. The animal is probably ailing. The first sign of illness is a reluctance to readily consume what they normally eat. This behavior is just as true for young calves on milk as for mature oxen on forage. When cattle go off feed that they normally consume, you have a problem.

If the problem is health-related the animal may appear listless and depressed. You must diagnose

the problem and treat the animal accordingly. If you are unable to diagnose or treat the problem, call a veterinarian. An animal that appears awkward, staggers, or in some other way does not seem normal might be poisoned or have some other disease problem. Prompt diagnosis and treatment are important.

Cattle are great at giving their owners messages about their feed. Be aware of the difference between reluctance to eat because your oxen are used to something better, and reluctance because of illness or hazardous feed or feeding areas.

Avoiding Waste

Like feeding other livestock, feeding oxen should be done in a least-cost manner. One of the simplest ways to minimize feed costs is to reduce feed waste. Feeding only recommended amounts and reducing parasite loads are the two easiest ways to minimize feed loss and keep working cattle healthy.

Parasite control begins with reducing exposure to parasites by feeding your cattle in clean uncontaminated areas and minimizing their exposure to infected animals or pasture. Separating new animals and treating them for parasites before they can infect your other animals is only logical. Treat your cattle for internal parasites before the pasture season and again a few weeks into the pasture season. Since young animals are more prone than older ones to problems from internal parasites, be especially vigilant about reducing their exposure to infection.

Many ox teamsters find that well-fed cattle not only get fat, but also waste a lot of feed. Except for young calves, working cattle do not need to be fed free-choice. To reduce feed wastage, feed your animals only what they need, by either:

★ meticulously following the feeding the guidelines for each steer or ox based on his age and size, or

★ reducing the amount you feed until your animals eat everything you put in front of them.

Feeding cattle in individual stalls in the barn is the best way to monitor feed intake and reduce feed waste. Offering hay in a rack and silage in a feedbunk will also help monitor and control feed waste. Properly designed feeding areas ensure that feed stays clean and is not wasted.

Most important, make sure your animals cannot step on the feed in their feeding areas. If cattle are fed hay on the ground, such as in an outdoor corral or pasture during winter months, try to ensure that they eat on clean ground. Spreading the hay in a new area each day not only reduces wastage (because they'll eat more hay if it is clean) but also reduces the build-up of manure in one area. Note that when hay is fed on the ground, cattle in groups of more than four or five will trample more hay than will a smaller group of animals.

Do not feed grain or silage on the ground because your animals may ingest contaminated soil as they try to lap up the grain and chopped feeds. Feeding grain on the ground almost always increases feed waste, as the grain gets stepped on, blown away, or washed away in spring rains.

Controlling rodents and birds helps reduce wasted feed. Besides transmitting disease, these pests both contaminate feed and consume a lot of it over time. Storing feed in closed, pest-proof bins will keep the feed clean and reduce feed loss.

Finally, cattle that are fed poor-quality hay or other poor-quality forages will waste more than if the feed is palatable and nutritious. Feed waste is expensive and should be avoided.

Oxen in Developing Countries

Oxen in Africa and some parts of Asia are often cared for much as cattle were maintained in early Europe or the Americas. They are expected to find their own feed and to survive under local conditions.

More work can be accomplished when the animals are fed and watered according to their requirements. Careful feeding is especially important when the animals are used for heavy plowing or other cultivation practices, which often come at the end of a dry season when the animals are in their poorest condition.

In Africa, oxen worked five or six hours a day have a hard time finding sufficient feed to meet their needs in the few hours they are grazed before being penned for the night. Animals that are corralled at night should be given feed and water in the corral. Bovines are designed to eat often, and in hot climates they eat more at night when the temperature is lower. Hay or green fodder eaten by animals while they are idle in the corral pays great dividends in the work they are willing to perform. Limiting water and feed stunts the growth of young animals and makes older animals more prone to health problems.

Lee Freeman
Thornton, New Hampshire

Lee attended an ox workshop held at the Remick Farm Museum in November 2005. An eager participant, he was interested in oxen but not quite ready to buy his first team.

At the workshop, Tim Huppe and I had 10 teams of steers or oxen and one single ox, all ranging in age from five months to 10 years old. Lee immediately took to the Milking Devon oxen, admiring their smaller size and vigor. As the weekend went along, and we worked oxen in many situations, his interest grew. Interacting with experienced and novice teamsters, he discovered that he enjoyed working cattle.

Not long after, Lee bought a pair of three-month-old Milking Devon bull calves, just weaned. He e-mailed us many questions about housing, fencing, and training. Lacking prior experience with cattle, he quickly learned that feeding newly weaned calves is not as easy as feeding mature oxen. In fact, he faced many of the challenges I mention when feeding young calves.

According to Lee, "I had quite a challenge feeding my steers Spencer and Clayton, from the time they arrived here at three months to the present (18 months). It was difficult to find general feeding guidelines. Having read as many books about steers, oxen, and cows, as I could get my hands on, I found there was no easy formula."

"When I picked the calves up," he went on, "the breeder/farmer gave me some quick 'care and feeding' instructions. Essentially, he said give them a coffee can of grain and all the hay they could eat.

"So this is what I did for the first three months. But after seeing photos of other breeds and visiting a few farms, I noticed my steers were significantly smaller than any other six- to seven-month-old calves. Also Clayton, my off steer, was getting quite a hay belly for a small steer. These two conditions — their overall size and the balloon belly — convinced me I needed some additional help. I then solicited help from you (Drew) and Tim Huppe. Essentially they delivered the same advice: too much hay, not enough grain, and check for parasites."

Lee then changed his feeding program. "With parasites eliminated, I quickly ramped up the steers to 6 pounds of 22 percent medicated calf grain and two flakes of hay. I then brought them up to 8 pounds of grain. This seemed to do the trick."

Looking back, Lee offers the following advice for novice teamsters with little cattle experience. "I think having a farmer mentor close by would have helped me through some challenges. . . . I have learned to listen and watch my steers to determine how much to feed them. If my wife or I get home for their afternoon feeding and they are wailing away, it might be time to increase their feed. Also, if they start pushing the limits of their fence, it might be time for an increase in hay." Finally, Lee points out after having one steer go off feed, "If they don't attack their grain, I look for reasons, and what conditions might have changed."

While there are many experienced ox teamsters and cattle owners who find feeding cattle easy, there are many first-time cattle owners who really struggle with young calves, as Lee did.

Now that Lee's steers are 18 months old, I would offer this advice. Do not fall prey to their wailing or bellowing for feed. This could now lead to overfeeding, which really limits the lifespan of working cattle. Cattle are quick to teach their owner when they should eat. Working steers and oxen are athletes, and as athletes, they need to be limit-fed once they are about 18 months old. At two to three years old, the steers will really start to put on weight on pasture alone. After all, it is their ability to graze that makes steers great beef animals and has allowed oxen to be the efficient work animals they are all over the world.

To keep a pair of oxen healthy and attractive, follow these five rules for feeding:

1. Feed each steer separately. By feeding individually, you can adjust each animal's diet for his own requirements. Every team has a dominant and a subordinate animal. If allowed to do so, the dominant animal will get the best feed and more of it. Each animal will get his fair share if you restrain or tie up the animals at least when they are fed grain or concentrates.

Under some management systems individual feeding may not be possible. When cattle are fed together as a group, invariably some animals in the herd are shortchanged. As a result they become stunted or malnourished. For a team to be in ideal condition and properly fed, individual feeding works best.

2. Make sure young animals get adequate protein, energy, and fiber. Feed the animals what they need. Since their needs change as they mature, closely monitor each animal's body condition, growth, and diet. Adjusting the diet of a young steer to ensure proper condition and growth is much easier than trying to change the animal after he has been stunted or grown

at a different rate from his mate. You can easily feed to grow two young animals of similar genetics, even if they are different sizes and conditions, if you control their diets and monitor their growth.

3. Avoid overfeeding mature oxen. Even for the teamster who is experienced with feeding cattle, learning to feed mature oxen is a challenge. We humans feel better when we have a few ribs showing. Your steers will, too. If you want your team to live for 15 or 18 years, keep them lean. If you want them to live only for 8 years feed them all they want. The choice is yours. Either way they have lived longer than most of their bovine brethren.

4. Watch your animals during feeding. Any deviation from their normal feeding behavior may be a sign that something is wrong.

5. Let your eye and common sense help you judge how much to feed. If your animals are thin, feed more. If they are fat, feed less. If they waste feed, you are wasting money. ★

4
PRINCIPLES
OF
TRAINING

Training oxen to do tasks is not a new concept. Oxen were exploited for work before humans used horses and other animals.

Cattle are intelligent animals. They learn something whether you are successful in training them or not. Too often the novice teamster ignores their natural desires and tendencies, leading to frustration with the animals and the training process. Many beginning teamsters give up training cattle because the task seems too daunting. Training oxen is not about magic. It is about understanding your animals.

Taking Time

I trained my first team under what I would consider less than ideal circumstances. I had no knowledge of training cattle, few written materials to fall back on, and only a basic understanding of bovine behavior. But as a boy I did have lots of time, patience, and some guidance from experienced ox teamsters.

Many years and many teams later, I know that those first steers were my best-trained team ever. My success with them was possible because the calves had been hand-fed from birth and were friendly. I was their provider and herd leader. I knew enough about cattle to make sure they never tested my dominance. In those days I had more free time than I have ever had since. That pair of Brown Swiss twins never learned that they could run away. They trusted my judgment the way a calf trusts a cow. Even when I made mistakes in their training, they never seemed to hold a grudge.

While not all teamsters have such a successful first experience, my goal is to encourage and inspire you to develop a team of cattle that are a joy to handle. Perhaps you have work that can be done with a team. Real work is the best training aid a teamster can use. A team of oxen learns best through repetition and consistent handling. Working a team in the field or forest after they have learned some basic commands benefits both the teamster and the animals.

Other than starting with young calves and being patient, consistent, and firm, no single technique works on all cattle. Cattle are individuals. Even the most experienced teamster encounters the occasional challenging animal or team.

All teamsters work and train their animals using slightly different techniques, cues, and methods. If possible, talk to successful teamsters about how they have dealt with the challenges of training cattle. Learn to compare and discuss methods of working cattle. Learn from the experience of others, and learn from their cattle.

Keys to Training

Training oxen is easiest when the animals are calves, but sometimes this is not possible. No matter what age they are, it is important to avoid discouraging the animals during training. A slightly tired animal pays attention; an exhausted and sore animal resists the entire training process. Learn to read your cattle and understand what you can expect of them.

ESTABLISHING RESPECT

Many teamsters fail in training cattle because they believe they should train their animals with kindness and adoration. As a teamster, you must develop a relationship with your animals based on respect and dominance. This relationship is easy to establish in younger animals that have been handled often, but it is difficult to instill in older cattle. Yet it is critical if you expect to have a team that can be trusted in the yoke.

Cattle are simple in their normal behavior, desiring feed, rest, water, shelter, and some social interaction. They are also herd animals that live by a hierarchy in the herd. You have to be the top animal in their herd. You can never let them know they have the power to dominate you. While they can learn to enjoy kindness and frequent attention from humans, enjoying such attention from animals not of their own species is not their normal behavior. Being fair and kind will aid you in training cattle, but will not train them. Even well-trained ox teams are content not to work at all, unlike some of our other domestic animals that look forward to daily workouts and training.

Expecting oxen to perform work under human direction for hours on end is a radical change in the adult bovine's normal behavioral pattern. As a teamster, you have to make the animal work. This may sound cruel, but once animals are trained they rarely resist working in the yoke.

Although calves are easily acclimated to this behavior modification, adult cattle often resist. Given the size and strength of these animals, an essential part of training involves knowing your

The Horns of the Ox

Cattle use their horns to show dominance. They will start as calves by rubbing their horns on you. As they get older, they use this as a way to test you and see your reaction. All cattle know exactly where the tips of their horns are and how to use them. Do not tolerate a calf, steer, or ox poking, pushing, or even "accidentally" hitting you. Younger cattle would never accidentally hit more dominant animals with their horns.

team. Before attempting to work with cattle of any age you must have a basic understanding of cattle behavior. And before attempting to train mature or wild cattle and to curb the undesired behaviors of these older animals, you must have a comprehensive understanding of cattle behavior and their natural tendency to establish their place in the herd.

A well-trained team of oxen is a joy to watch and work.

CATTLE BEHAVIOR

Like most other farm animals, cattle are gregarious, social animals that tend to seek out the companionship of members of the same species. In fact, an animal that has belonged to a group or herd all his life becomes seriously distressed if he is removed from his companions. Being social animals, cattle are able to bond to a new group only after they establish their place in the new herd. This aspect of dominance and subordination is the natural way for cattle to interact and is important to understand if you are going to train cattle.

In addition to their social behaviors, cattle have specific feeding and resting behaviors. They also have numerous sexual behaviors (which for the most part are eliminated by castration).

Training and working oxen takes time away from these normal routines of feeding, drinking, ruminating, resting, and social interaction. This time away from normal behaviors can stress the animal being trained. Add a recent weaning, separation from the herd, harsh treatment during handling, and too

The younger cattle are when training begins, the easier it is for animals and teamster to learn to work together.

many demands early in training, and you invite susceptibility to disease and some bizarre behavioral changes.

For older animals, the early period of training is a critical time of acclimation and learning who is boss. An animal that is poorly handled at this stage may similarly become sick or display behaviors such as lying down in the yoke or allowing himself to be dragged. He may also fight the entire process. Do not let him fight you. Plan ahead and anticipate.

WORKING IN PAIRS

Working with two animals that are familiar with each other can be easier than working a single animal. The pair will be easier to control in the yoke, and the presence of a teammate has a calming effect. It is a good idea to use a well-trained animal to help calm and train a novice one, provided there is no chance that the latter will challenge the well-trained ox.

An older, well-trained team that has spent considerable time in the yoke will rarely be separated by choice. Where you find one ox, the other will be nearby. This social behavior has advantages when you try to find a team that has been released to graze. In fact, if something dire happens to one ox in a team, some teamsters advise getting rid of the remaining teammate.

A well-trained ox, however, is the result of a lot of time, experience, and expense. Getting the ox to adapt to a new mate is easier and faster than starting a new green team.

DOMINANCE AND TRAINING

In every team one animal is dominant. This animal will get the best feed, drink first at a watering space with limited headroom, and to some degree dominate the other animal as the pair move about their living space.

Cattle establish their dominance by force. Body weight and horn size play an important role. Again realize you have neither horns nor size to dominate oxen, so you have to figure out how you will become the dominant animal through the advantage of greater intellect.

Even though the relationship is established through force and physical aggression, maintaining the relationship requires far less energy. Most subordinate animals retreat at the slightest sign of aggression from the dominant animal. Animals will occasionally spar as if playing, and the subordinate animal may take advantage of newfound size, strength, or horns to gain status in the herd.

Social behavior and dominance are key elements of working and training cattle. To train cattle effectively you have to physically prove to your animals that you are dominant. The dominance must be complete, much like the dominance a large bull has over a herd of smaller cows. Calves are easily intimidated. Establishing your dominance over calves is therefore easily accomplished.

In mature cattle that have had little or no handling, establishment of dominance is the biggest challenge to training. It must happen quickly

Natural Leaders?

In the natural herd environment, the dominant animal is not always the leader. Usually the largest bulls follow the cows. When they have reached their destination, the bulls push their way to the front to get the choicest feed or water. The leader may be the most experienced animal or an older animal of low status. The herd recognizes that experience and follows eagerly. This helps to explain why oxen are so easily trained.

and effectively or training will be difficult, if not impossible. If an animal can prove himself more dominant or powerful than his trainer, the animal will be dangerous to work with.

Using Your Advantages

Like the dominant animal in the herd, you must convince the subordinate animals that they may never challenge you. Establishing this relationship takes skill, thought, and mechanical advantage. Even a small calf can overpower a large man who is not prepared to catch and restrain the animal. A properly captured and restrained animal learns to

Cattle establish their dominance by force.

respect the trainer. Maintenance of the relationship is easy compared to the initial capture and restraint.

As a teamster you must not only be dominant, but you must also prove yourself worthy of being followed as a leader the cattle can trust. Cattle require adequate feed, water, rest, shelter, and protection from danger and injury. They will be easier to work when they realize that you provide them with these basic needs.

Background and Experience

Genetic and environmental factors may have an impact on how long the initial dominance training takes. An animal that has at one time proven his strength, skill, or speed to outmaneuver predators or people is going to be hard to control. Don't underestimate an animal's strength or experience, especially if he is mature. The bovine that has come to rely on his own skills or instincts will be a challenge for even the most competent teamster. Be patient. Cattle learn only if they are treated fairly and never allowed to realize the immense size and power advantage they have over their trainer.

An elephant trainer once told me that you could train an elephant quickly by chaining it down and beating it to prove your dominance. Such an elephant will work as long as he thinks he might be beaten. However, when the opportunity arises for the elephant to realize he can dominate the master, such training will come back to haunt the trainer.

. . . .

Cattle learn only if they are treated fairly and never allowed to realize the immense size and power advantage they have over their trainer.

. . . .

A better technique is to start the animals young, work with them, respect them, and not force them brutally to work. In the end, the ox or the elephant that has learned his place through patience and repetition will be a much more enjoyable and trustworthy animal to work than one that works out of fear or due to severe physical restraint.

BREED CHARACTERISTICS

Most teamsters and cattleman agree that there are differences in working with different breeds of cattle under the same management and handling, no matter when the training begins. Some breeds are always fast-moving and eager to outmaneuver their teamster, while others are content to work placidly

Animals with similar genetics, handling, and training may nonetheless behave as two distinct individuals in the yoke.

A nineteenth-century English cattleman named Adams once stated,

"Where the ground is not too heavy, the Devonshire oxen are unrivaled at the plow. They have a quickness of action that no other breed can equal and few horses can exceed. They are sometimes trotted along with empty wagons at 6 miles per hour, a degree of speed no other ox has been able to stand."

in the yoke. The strong and weak points of the breed you choose to work with will affect your ability to train your steers.

Two extreme examples of breeds commonly used as oxen in the United States are the Brown Swiss and the American Milking Devon. Both have a history of being used for draft, milk, and meat. Under identical management the two breeds will behave differently in the yoke.

As a rule the Brown Swiss is quiet, docile, slow, moving, and easily handled. This quietness is carried to such an extreme that some individuals are sluggish and tend toward stubbornness. My first team were Brown Swiss. Their calmness was an advantage for me as a novice teamster. The slow deliberate way this breed moves can prove safe and enjoyable for the beginning teamster.

By contrast, the American Milking Devon is a fast-moving, active animal well noted for its intelligence and ability to outsmart a young teamster. Some teamsters go so far as to claim these oxen have to be retrained every time they are put in the yoke. Others feel the extra effort required to train them is not worth the reward. Nevertheless, people who rely on oxen to make a living have long cherished the characteristics of Milking Devons. According to some, Devons in the yoke are second to none.

No one breed of cattle makes perfect oxen. A breed that is easily trained or gets extremely large may not be appropriate for a particular teamster. Choose your cattle based on how they will function in the yoke and your own preferences in training and handling.

Experienced teamsters who have worked with a variety of breeds say that all breeds train easily if they are handled when young and begin their training early. Besides, even animals with similar genetics,

handling, and training may behave as two distinct individuals in the yoke. This individuality is one of the challenges and joys of working with oxen.

FLIGHT DISTANCE

One of the biggest hurdles in working with cattle is overcoming the animals' flight distance, or the point to which you can approach the animal before he turns to move away. A well-trained team should be approachable at anytime, anywhere. Whether or not the team is in the yoke, the teamster should be able to readily touch and catch the animals.

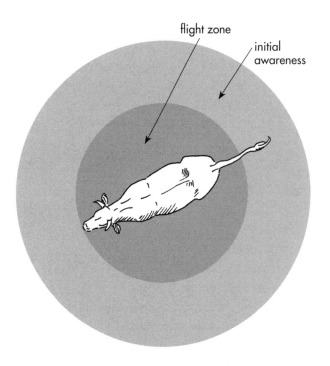

flight zone

initial awareness

An animal's flight distance — how close you can get before he turns and runs — indicates the ease with which the animal may be trained.

Most animals have a natural fear of the unknown. When you approach an untrained animal you will most likely observe an initial awareness. He will direct his head toward you as you approach. Soon he will orient his eyes, nose, ears, and then entire body toward you. As you get closer he will enter a state of alarm or fear. Get a few feet closer and the animal will turn and run: this is the flight distance.

In tame cattle this point is closer than the flight distance of untrained animals. Well-trained oxen have *no* flight distance. They come when they are called and do not run or show alarm when you approach.

An animal's flight distance tells you how to handle him. Some cattle are so wild they run away as soon as they detect the presence of a human. Such cattle are the most difficult to train. They have developed a fear of humans and their training will take a long time. They may never work willingly. Animals that allow people to come within a few hundred yards may be better, but animals that are used to being herded, touched, or led will be the easiest animals to train and work.

Easiest of all to train are young calves that have been handled since birth, and especially those that have been hand- or bottle-fed. Once these animals are introduced to the basic commands needed in the yoke, and are led around individually or in pairs, their flight distance is minimal. If they are consistently trained and frequently handled when young they will remember the experience for a long time and be more easily put to work later in their lives.

AGGRESSIVE BEHAVIOR

Cattle are not afraid to use force if they feel they are in danger and cannot retreat or if they are protecting their young. Many halter-trained beef cows have surprised their owners at calving with newfound aggressiveness. Never underestimate the aggressiveness of cattle that have never been handled. While they usually retreat from people, they more readily stand their ground than do sheep, goats, horses, or other farm stock.

Most trained oxen are calm, easy-going, and forgiving beasts. Even suffering cruelty and under the worst circumstances, most will never attack the teamster. They will instead retreat or, if they feel escape is not an option, will endure the circumstances.

Oxen that are not aggressive toward their teamster may be aggressive toward other people, especially children, if they think they can dominate them. Given the chance, your cattle may establish dominance over someone without your ever knowing it. Oxen should learn to respect all people and view them all as dominant.

Cattle that succeed in bluffing people into retreat or chasing people out of their living space have learned they can dominate people. Be careful not to allow your team to be tormented or teased by anyone. Even the well-trained team may easily cause injury or worse to an unsuspecting person if they are allowed to think that they can dominate humans.

Sensory Awareness

Creatures of habit, cattle are content to do the same thing day after day, provided they are not uncomfortable. Given the chance to do something new or more enjoyable, however, they will quickly leave the normal routine. Enjoyable activities include going out to pasture in spring, leaving the barn after months of being indoors, and getting fresh bedding.

Given the opportunity, experienced animals will normally lead the way with lots of excitement, and the younger ones will follow. The inexperienced animals usually investigate any new event or object by sight and sound, then smell, and eventually taste. Over time, cattle can become habituated to many noises, smells, and sights that other draft animals would shy from.

Even when cattle seem to be peacefully grazing, resting, or ruminating they are aware of their environment and any changes in it. This awareness comes by means of one or more of their senses.

SIGHT

With their eyes located on the sides of their heads, cattle are able to see all the way around themselves in a panoramic view. They don't see colors the way people do, which helps them easily distinguish movements. In the wild this ability helps them to see approaching danger. In the yoke it allows them to watch all the teamster's actions.

Since oxen are usually worked in a yoke, both the yoke and the teammate may block normal vision, but the animals soon learn to turn their heads to see what they need to. Since cattle watch the teamster at all levels of training, visual cues are important. Most animals respond as much to consistent visual cues as they do to voice commands. Teamsters who are inconsistent about matching their movements to voice commands have a hard time controlling their teams.

Oxen are often far more observant of us than we are of them. Your cattle will notice the way you walk and move, and they will respond accordingly. If you approach them quickly, with force and aggressiveness, they usually retreat. They are visually reading your mood. If you raise the whip faster than normal, they may detect a coming hit. Such a cue will cause a well-trained team to stop acting up.

Cattle may be driven with largely visual and only occasional physical cues. Large teams in pulling contests have been handled successfully by mute teamsters. Even teamsters who believe that their oxen drive only with voice commands would be surprised to learn how much their team responds to subtle visual cues.

SOUND

Cattle have good hearing, making it pointless to yell or talk loudly to your team. If they are distracted or are pulling an implement that makes a lot of noise you may need to be loud to get their attention. Many an ox, however, has fooled his teamster into thinking he isn't listening.

Most commands are accompanied by visual or physical clues, but a well-trained team will respond to all of the commands given by voice alone. Train your cattle in your own native tongue, but make sure your voice commands are different enough to avoid confusion. Even when the words are similar, most cattle will understand what you expect if you give the commands in different tones.

Since cattle cannot understand spoken language, some teamsters believe you should not speak to your team unless giving a command. In this way the team learns to pay attention when you address them. A preparatory cue may be a whistle, a click, or some other sound that signals to the cattle that they are to pay attention.

If your team does not do what you desire, but you continue to give voice commands, you may teach them to respond with the undesired behavior. An example is a team that learns to shake its head when you say "back." If the team resists backing and you are hard on the animals for resisting, they learn to associate the cue "back up" with getting hit on the head, which teaches them to shake their heads in resistance. This habit can be so bad that a poorly trained team may never obey the command to back up.

An ox responds only to the teamster's actions. If you give inconsistent visual, physical, or voice cues, the animal's training will take longer and be more difficult.

TOUCH

Cattle have a keen sense of touch, even though their hides are tremendously strong. When insects and other parasites irritate them, it becomes obvious that certain parts of their bodies are more sensitive

In an 1849 letter to *New England Farmer* a Vermont farmer named Reynolds claimed:

OXEN LORE
"I think many teamsters talk too loud to their oxen. They hear as quick as a man when called for their food, which proves that their hearing is good. Why then should a teamster scream so hard as to be heard a half mile or more?"

Les Barden, an accomplished teamster from Rochester, New Hampshire, states:

"The habit of continually talking to cattle serves no valuable purpose. More than likely it only serves to vent the nervousness of the novice teamster. Before every voice command the animal should be given a preparatory command (but do not use the name, which should be reserved for speaking to an individual animal) in order for the animals to realize they must pay attention."

than others. The slightest touch of a thread to the hide at the shoulder or over the ribs can cause an animal to twitch his skin to rid himself of a pest.

Teamster's Dictionary

Cattle cannot understand words. Words are just cues you have repeated when your animals did what you desired. Cattle can learn to distinguish among a number of verbal commands, including their names. Many New England teamsters have cattle that respond to more than 10 different verbal commands.

Aside from responding to their own names, the commands cattle learn to obey include but are not limited to:

1. Get Up = move forward

2. Easy = slow down

3. Whoa = stop

4. Gee = turn right

5. Haw = turn left

6. Step in = step toward a pole/chain

7. Step out = step away from a pole/chain

8. Back = move in reverse

9. Head up = lift their heads when in yoke

10. Come boss = come in from pasture

At the other extreme is the ruggedness of the horns and head. In battle, cattle push and shove with their heads, withstanding tremendous force and pressure.

These extremes of sensitivity help us understand how little it might take to get an animal's attention, compared to how much bodily force the animal could exert with little or no harm to itself. Cattle are rugged animals that can withstand tremendous strain, but that does not mean they should be subjected to pain. As important as it is to show your dominance over your animals early in training, you should use the whip or goad only as a physical cue or reminder.

Negative reinforcement with the whip may be necessary, but use the whip only in a controlled and well-directed manner. Cattle are easily confused, upset, and frightened. When they do not understand why they are being hit, wild cattle can sustain a tremendous whipping without learning anything.

Just as the mosquito or fly causes an immediate and predicable reaction when it lands on an ox, the whip or goad should always cause an immediate and predicable reaction that you control. The touch of the whip should be well-directed, controlled, and forceful enough to get the desired response. Use the whip as an extension of your hand to get the animal's attention or to correct a wrongdoing.

Senseless beating confuses cattle and causes them to regress in their training. A physical cue should always be used in conjunction with a preceding visual or verbal cue so the animal will be conditioned to react before the whip strikes him.

When cattle are being handled for the first time, they must be restrained in such a way that they cannot escape — with no injury to the animal. The capture must be as pleasant as possible. If the initial capture is painful, or if it precedes minor surgery or injections, the animal immediately associate capture and handling with fear and pain. Cattle quickly learn to avoid anything they find unpleasant.

Intelligence

Teamsters have debated for centuries about the ox's intelligence. Are cattle smart enough to deserve a discussion of their intelligence? Is an intelligent ox one that always behaves, even under the cruelest treatment? Or is an intelligent ox the one that refuses to work, balks under poor handling, and runs when he knows he can get away?

Cattle are not stupid. In maze tests they have performed remarkably well, even better than swine or dogs when vision alone was needed to solve the maze. In fact, they often train their teamsters to react to their cues, rather than the other way around. Their intelligence and capacity for learning have fooled many people into believing they are too dumb to work. Teamsters who prefer animals that are not so hard to train often discard both the American Milking Devon and the Chianina.

It is not the dumb animals, however, that resist training. Those that resist working or choose to work at their own pace are likely to be the more intelligent bovines. Oxen's intelligence, combined with their rebellious behavior, has caused many a potential teamster to give up training the animals.

As a teamster you have to understand the intelligence of your cattle and anticipate what they will do before they do it. We humans possess a degree of intelligence that is far superior to that of the ox, but poor preparation on our part to counter their smaller intellect may lead to failure in training them. Whether you handle an old experienced ox or an animal that has never before been handled, anticipating the bovine's movement is the best way to exercise some control over his behavior.

HOW CATTLE LEARN

Learning might be defined as a relatively permanent change over time, in response to practice or patience. Cattle probably learn far more than we give them credit for, and they do this in several ways.

Imprinting

Soon after birth a calf imprints on his dam, the herd, and his immediate environment. A teamster can use this period of imprinting to ensure that the calf does not associate humans with any threat or

Use the whip or goad in a controlled and well-directed manner.

William Youatt, a nineteenth-century English cattleman, spoke favorably on the intelligence of the ox:

"Cattle, like other animals, are creatures of circumstance. We educate them to give us milk, fat, and flesh. There is not much intelligence required for these purposes, but when we press the ox into our immediate services, to draw our cart and plow our land, he rapidly improves upon us. He is in fact an altogether different animal. When he receives a kind of culture at our hands, he seems to be enlightened with a ray of human reason, and warmed with a degree of human affection. Many dairy and beef cattle have just enough wit to find their way to and from the pasture, but the ox rivals the horse in docility and activity, and fairly beats him out in the field of stoutness and honesty of work."

danger and that the animal becomes comfortable with being handled. Some of the most easily trained cattle are those that did not imprint on their dams at all but on humans instead.

Habituation

Animals (and humans) develop a decreased response to stimuli that they experience repeatedly. A team that is hitched to a load for the first time, for exam-

A team in the yoke must understand that they have no choice but to follow your direction. If the animals once get their own way, training will be more difficult.

ple, may try to run from the load. After a few tries the animals become accustomed to the load and realize it will not hurt them. This same mode of learning, called habituation, takes place when cattle are first captured. If they learn that capture will not be painful, they are habituated to the experience.

Imitation

Imitation is learning by following example. Calves learn what to eat by imitating their mothers. They follow the herd, learning the normal routine over time. While cattle cannot train other cattle to work, an untrained animal yoked with one that is trained will learn through imitation. The animal eventually learns what he should do in the yoke by following the trained ox.

Association

Also called "operant conditioning," teaching by association is one of the most successful techniques for training cattle. This technique involves reinforcement that is dependent on a voluntary response from the animal. You give the animal voice, visual, or physical cues and he voluntarily responds. The reinforcement may be positive or negative. The fastest and most effective way to teach cattle is to administer the reinforcement as soon as possible after the desired or undesired response. Delaying reinforcement leads to frustration on the part of both the animal and the trainer.

Cattle that have had a bad experience, such as while hitched to an implement, may defecate as they approach that implement. Defecation in such a situation is a sign of nervousness. The team has been conditioned, probably unintentionally, to associate the implement with fear.

LEARNING BAD HABITS

Your cattle will learn, no matter how successful or unsuccessful you are as a teamster. Like all animals they can learn bad habits as well as good ones. They may learn how to do what they want to do just as easily as they learn how to do what you want them to do. They are not dumb beasts if they have learned how to make you upset enough to take off their yoke and let them return to the barn or pasture. In such a case you have conditioned the animals through positive reinforcement. When the animals act up, they are rewarded by being allowed to do what they want, which is to graze peacefully or chew their cuds in the shade.

Cattle in the yoke must always do everything you want them to do. If they figure out just once that they can act up and get away with it, their training will become more complicated. The most difficult behavioral problems to eliminate are self-reinforcing ones.

Your challenge as a teamster is that cattle quickly learn routines, such as where they are yoked and unyoked, where they are fed, and where to turn to go home. Many teamsters unwittingly give their animals hints when a workout is ending. Be aware of the subtle cues that reveal your intentions. Vary the training area, yoking area, and pathways so your team never knows when the training or workout will end.

Handle Them Often

Frequent handling is the key to training cattle. The more time you spend with the animals the more successful the training will be. They have to learn to understand your cues and be willing to follow your directions. In contrast, if you yoke them infrequently, you will end up always having to force a steer or ox to do what you want, when he has no interest in following your lead.

Some of the most easily trained steers imprinted on humans, rather on their dams.

OXEN
LORE

R.E. Pike in *Tall Trees, Tough Men* describes the historical use of oxen for logging:

"When an ox couldn't pull the load, and the driver wouldn't give up, the bellowing of the bulls [oxen] would even make a strong man cringe."

Your animals must not disobey or shy from you at any time, even if their workout is over. If your cattle act up and think it's time to go home, hitch them back up and continue working until they are sufficiently tired not to turn toward home every time you pass the homeward path. Cattle must learn that when they are in the yoke they must always be responsive to you, their teamster.

Cattle are intelligent beasts that will do their own thing if given the opportunity. When trained properly and not given the chance to outsmart the teamster, their intelligence can be successfully employed to get a lot of work done.

Whether previously trained or not, cattle watch people for cues as to what to do; your actions as their trainer have a direct bearing on how easy or difficult the training will be.

CHAPTER ROUND-UP

Training a team of oxen has been called an art. Much as an artist shapes a stone or log into a statue, the teamster shapes two independent and rebellious animals into a team that willingly responds to commands. This shaping takes an understanding of the animals and the tools at your disposal. It also means that there are different ways to achieve the same objective. I have observed many different training methods around the world.

Compared to shaping wood or stone, the challenge in working with a living team is that the animals have already been shaped by their prior exposure and experience. The cattle will learn something every time they are in the yoke. A stone or log will never challenge the artist the way that living animals, with a keen awareness of their needs and surroundings, can and will take advantage of the teamster. Compared to the development of other artistic skills, the training of oxen is in some ways more challenging.

Training cattle is more than executing certain steps and exercising skills. It is a combination of prior knowledge, psychological dominance, physical dexterity, and repetition of verbal and visual commands. Preparing yourself to train a team includes knowing your animals, modifying their environment to make training possible, and understanding the rules of dominance that are genetically programmed into the ox's small brain. ★

Yoking a Team

Before yoking an ox for the first time, be sure the animal is securely tied.

1. Hold the bow in place on the off steer's neck.

2. Place the bow through the holes in the yoke.

3. Add spacers to maintain the bow depth, and place the bow pin through the hole in the bow.

Step 1

Step 2

Step 3

4. Bring the nigh steer up and hold the bow in place under his neck.

5. Raise the yoke with one hand while bringing the nigh steer under the yoke, then lower the yoke onto the nigh steer's neck.

6. Slide spacers onto the nigh steer's bow as necessary and fasten the pin. Only one pin per bow is needed.

Step 4

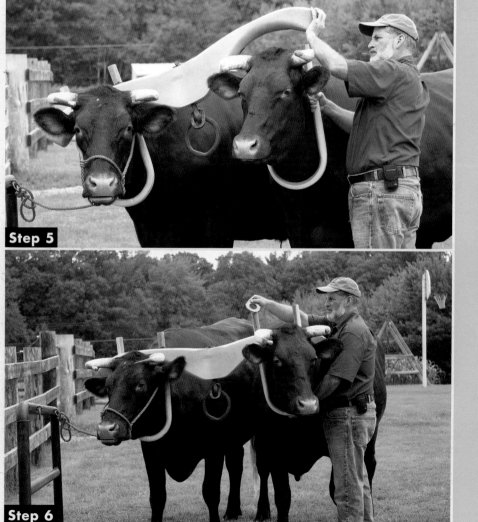

Step 5

Step 6

Yoking Calves

1. Place a bow over the neck of each animal so it will be readily available for the next few steps.

2. Place the yoke over the calves' necks.

3. Turn the bows over and fit them through the holes in the yoke.

4. A close-up of the bow spacer and bow pin.

5. Add spacers as needed and push the bow pin through.

Step 1

Step 2

Step 3

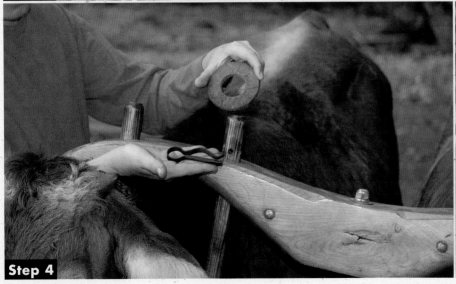

Step 4

Step 5

Hitching

Bring the team around to the front of the pole or tongue on the wagon. It's usually easiest to approach from the left to hitch the team.

1. Step the off steer over the pole, and then step the nigh steer in, toward the pole, so he is parallel to the pole and the off steer.

2. The team is ready to be hitched.

3. The pole or tongue on the wagon is pushed through the yoke ring and then held in place with a T-pin.

Step 1

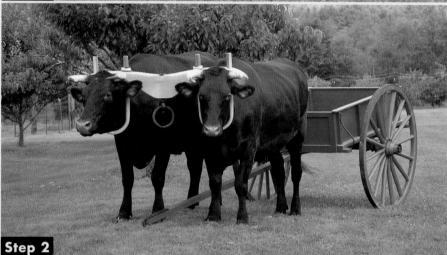

Step 2

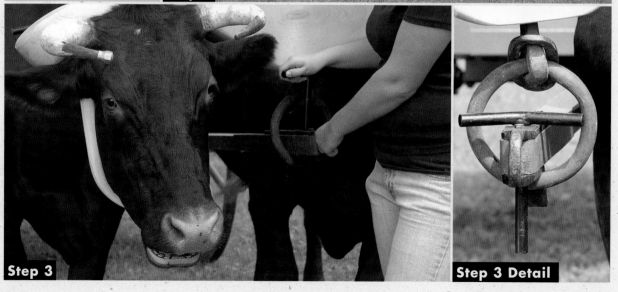

Step 3

Step 3 Detail

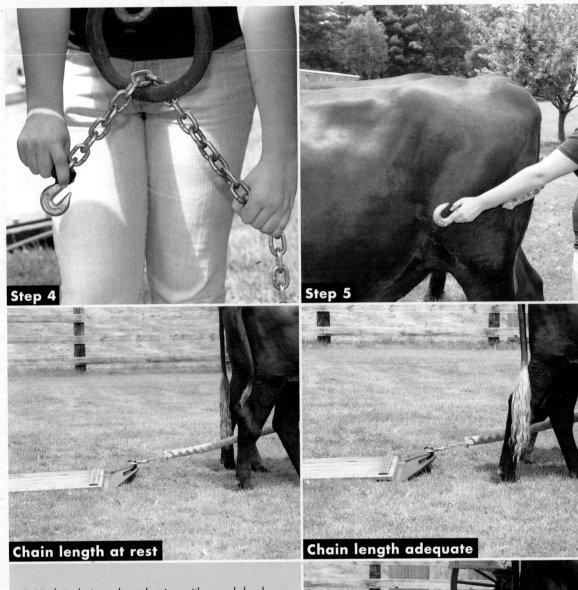

Step 4

Step 5

Chain length at rest

Chain length adequate

Minimum chain length

4. Hitch a chain to the yoke ring with a grab hook on the end of the chain. Then hitch the grab hook back onto the chain.

5. Estimating the chain length is critical. The chain should be neither too long nor too short. Some teamsters, as demonstrated, will pull the chain back to the steer's flank. This will provide adequate chain length for the steers' back feet to clear the sled or stoneboat, as shown above. This does provide more chain length than is usually necessary, depending on the hitch point on the sled or stoneboat. The photo to the right shows the minimum chain length, with the foot just clearing the sled on a forward pull.

Pulling a Load

Step 1

Step 2

Step 3

1. Both animals in the team must be trained to lean into the yoke together and pull with their heads up, slightly below an imaginary line off the top of their backs. The teamster stands beside the team, in a position that will able him to direct the animals in a turn or stop them. Provided the yoke fits the team, they should hold their heads as shown during a pull. If one animal drops its head or throws its head up, the bows need to be evaluated for proper fit.

In a heavy pull, such as in competition, the oxen will learn to push up as well as forward in the yoke. The animals are trained to do this to maximize their power and minimize the drag on the front of the stoneboat.

2. The logging cart or a forecart has a higher hitch point than a sled or stoneboat, but the same head position is desired, during a heavy pull. In this photo the off steer is lagging a bit, so the teamster will tap him with the goad stick, in order to keep the team working together.

3. This front view of the team and teamster, pulling a load, shows the comfort of the steers, as determined by their head position. The teamster can correct a steer that either speeds up or slows down while pulling.

Training to Lead

1. Initially it is best to capture the steers with a bit of grain and then tie them up in the barn, on a hitching rail, or to a chain and hitch point in the field (as shown here). Once caught and tied up, it is easier to lead a calf with a halter than with a collar and lead rope.

2. Once the halter is securely placed on the animal as shown, pull the calf forward. If he is initially reluctant to lead, have someone behind the animal tap his rump or push him along. Do not drag the calf, but take advantage of his natural instincts to get away. Step behind the head if necessary to get him moving forward.

Step 1

Step 2

When tying a calf, make sure he is tied to something he will not break, with a quick-release knot. When he does what you ask of him reward the calf for good behavior with a pat or rub on the neck.

Howard VanOrd
Russell, Pennsylvania

A stunning photo of one of Howard's oxen appeared on the front page of *Small Farmer's Journal* in the mid-1980s. Taken by Gary Lester at the Warren County Fair, it showed a huge ox named Danil beside a young "Jay Jay" Frentze. Beautifully trained, wearing no rope or halter, Danil stood amid the fairgrounds with only a young boy, casually holding a whip, at his side.

During 1985 and 1986 I had been trying without much luck to take a photo for the cover of *The Oxen Handbook*. I remembered Gary's as one of the most eye-catching photographs of oxen I had ever seen. Thanks to Lynn Miller (editor of *SFJ*), my then-publisher Doug Butler, and Howard, we were able to use the photo on the cover of my first book.

Howard is a true artist when it comes to training oxen. He knows oxen, he knows how to train them, and best of all he knows how to help other people with their oxen. He is also one heck of a good storyteller.

When asked how he got started with oxen, Howard replied, "In 1957 I was working for the County Home and Farm. We had a pair of twin Hereford-Holstein calves born. The boss gave me permission to train them for oxen. At the time, there was a patient/resident at the home named 'Pop' Weir. For years, he had been the blacksmith at a logging camp where they used oxen. Pop gave me my basics on working oxen."

He went on, "I later became involved with an old-timer, Irving "Johnny" Lamb from Friendship, New York, who seemed to know everything about oxen and was more than willing to share his knowledge." Then, "in the late '70s, Ray Ludwig was a big influence and help when I started the first oxen 4-H club in Pennsylvania."

Howard's ability with cattle may seem like magic, but he will be the first to tell you otherwise.

"In my opinion," he says, "the basic keys to raising oxen are to be consistent in your commands and to make them obey your commands. There has to be mutual respect. Always remember that anytime the ox can see you or hear you, you are training him. You and the ox are in training anytime you are together. I have learned that oxen will take on the personality of the owner or trainer and if you don't show them respect, they won't show you respect. I also learned that the cow will make just as good a worker as the steer will."

With 50 years of experience with oxen, Howard never kept an exact count but guesses that he's had 25 or 30 teams. Among them were Holstein-Herefords, Holsteins, Brown Swiss, Ayrshires, Brown Swiss–Holsteins, Linebacks, Devon-Ayrshires, Holstein-Chianinas, Milking Shorthorns, Devon-Holsteins, Devon-Normandys, Charolais, and, quoting Howard, "the very best breed, *Devons!*"

His best advice, in one sentence, is: "Make sure they know you are the dominant member of the team and you show them respect, kindness, and patience."

5
TRAINING STEERS

The desired result of training steers is to have a team with spirit that works eagerly for you. Getting to that point requires developing a specific plan that you will follow throughout the training. The plan need not be written or elaborate, but should follow a flow of events that you control. You should not react to the behavior of your cattle, but rather the animals should react to your training techniques. In New England, eight-year-old teamsters have successfully trained oxen for hundreds of years using little more than common sense and commitment to the endeavor.

First Steps

The initial steps in training are to evaluate the animals and their behavior, then select an appropriate method of capture and restraint. This first encounter is critical to the success of your entire training program. Training begins the moment a team of cattle has their first contact with humans. That first experience affects the animals in some way, whether good or not. During the first meeting you must do nothing to frighten or hurt your calves. They must learn to trust you for their basic needs and not associate humans with pain or fear. Even with cattle that are accustomed to being around people, the first steps in training should be slow and deliberate.

Once your team are named (see box), it is ready to begin training. How you approach the initial training depends on the degree of socialization your animals have had with humans. As the teamster you must establish dominance early so your animals will never realize their strength and power.

Dominance is not cruelty. Cattle rarely subject each other to undue pain and suffering, even when one animal is dominant over another. Too much force may cause your team to lose their spirit and become depressed and lethargic in the yoke. Use only enough force to control an animal and his movements.

INDIVIDUALITY

Just as every human is unique, so is every ox: no two animals are exactly alike in behavior and temperament. Despite their simple behavior and needs, cattle can be difficult to understand. Even steers

Naming Your Oxen

Give a lot of thought to the names you assign your animals. A pair of oxen should have names that:

- contain one syllable

- do not sound alike

- do not sound like any of the commands

Buck and Jim, for example, make a good pair of names. Bob and Hob sound too much alike, leading to potential confusion regarding which animal you are addressing. Maw and Paw not only sound alike, but also may easily be confused with the "haw" command.

with the same parents and training may behave differently in the yoke. Individual animals respond to commands, to handling, and to the tasks you ask of them in various ways.

While some steers are amazingly easy to train and eager to work, others refuse to work adequately to ever achieve a high level of training. In the northeastern United States top teamsters in pulling competitions cull a lot of cattle before finding two animals to make a competitive team.

The ideal situation is to have two animals that behave and act similarly in the yoke. Two fast steers or two slow steers can be worked together, but one fast eager ox and one slow lazy animal will have difficulty as a team.

According to Frank Scruton, an accomplished ox pulling competitor:

"You ought to know in the first month of training whether the animals are good enough to keep. Some animals just don't have what it takes to pull."

Although sometimes the best solution is to cull an inferior animal before investing too much time in training, don't be too quick to discard animals from your first team. A beginning teamster can learn a lot from an ox that is not perfect.

MAKING A TEAM

The temperament, size, or behavior of an ox dictates on which side it should be placed in the yoke. Some teamsters place the fast-moving, hard to control ox on the nigh (left) side for better control. Others place the slower animal on the nigh side to better keep after him and coax him along.

Many teamsters put the larger, more powerful ox on the off (right) side so the team always pushes toward the teamster, or to compensate for the uneven yoke when the off animal is in the furrow, during plowing. The smaller, less powerful steer is then on the nigh side where the teamster can give him more coaxing and can more easily see the larger animal to the right.

An experienced teamster quickly figures out where each animal should be placed in the yoke. Training is simplified when each animal is assigned to one side and kept there. If problems arise, bad habits develop, or a teammate dies, however, an animal may be retrained to the opposite side.

Six Rules of Training

Experienced teamsters follow six rules that are essential to proper training. Failure to follow just one of these rules may lead to a team that cannot be handled, is untrustworthy, and, worst of all, may be dangerous. The wildest and most unruly team, however, will eventually respond to these six rules of training:

1. **Set goals.** Have a specific plan for training your team. Oxen can learn only a few commands at a time. Decide which commands you will teach your team first. When they master those commands move on to the next set of commands. Know and understand the restraint and reinforcement systems you will use and, more important, when you will use them.

   As part of your goal-setting, plan a schedule for working your team each day or week. A team needs regular work and training. The problems experienced by many beginning teamsters stem solely from spending too little time working with and learning with the animals.

2. **Maintain control.** Your skills and techniques as a teamster will have a direct

Calves may be trained by any child with a basic understanding of the rules of training.

impact on how well your team reacts to the commands. Think about how you will maintain control over your animals. When they misbehave, losing your temper will likely lead to inconsistent commands. You cannot physically restrain animals that are many times your size. You must be able to gain and maintain full control of the animals at all times. Maintaining control usually requires some mechanical advantage, such as a halter and a lead rope.

When things go wrong, be ready to blame yourself and not the animals. Teamsters who cannot control their teams usually bring the situation upon themselves. At every stage of training be prepared to control your team using the most effective means necessary. Realize that properly trained oxen can eventually be controlled without lines, nose rings, halters, or ropes.

3. **Be firm.** To gain and maintain dominance over your animals, infuse every command and action with authority.

4. **Be patient.** Cattle can learn only a few things at a time. Just getting one animal to follow you willingly may take days or weeks. This slowness to learn is not a result of stupidity. It is a result of intelligent resistance to something the steer is unfamiliar with.

5. **Be consistent.** Cattle watch and listen even when you are not speaking to them.

Inconsistent signals, cues, or behavior on your part only confuse your animals.

6. **Work regularly.** Any animal left idle becomes more difficult than one worked regularly. Farmers of old who had endless work for their draft animals once prized a team full of energy. The best way to use that energy was to creatively work it out of them.

Today many teamsters in the United States yoke their teams a few times a year for show and expect perfect behavior. Other teamsters brag about how they haven't yoked a team in more than a year. Such animals, even though beautifully trained, are often individuals or breeds that have been selected for their large size and lethargy. They have not been selected to work at a fast pace. Regular human contact and big feed buckets have conditioned them to walk slowly in the yoke. Such animals would be shocked if asked to do more than walk or stand in a yoke.

Old-time New England teamsters felt that six thin, rangy, and willing steers were better than two big fat lazy steers consuming the same amount of feed. You can't help wondering how today's large lazy animals would be viewed in cultures where cattle have to earn their keep as draft animals.

Real work is an excellent training aid, acting as a restraint while continuously exposing the team and teamster to new challenges.

Ten-Step Training Schedule

Here is an example of a step-by-step training plan you can use and adapt to your own situation.

Step 1: Leading. Start by training your animals individually to lead with a halter.

Step 2: Handling. Accustom your animals to regular handling with frequent tying and grooming.

Step 3: Introduce the yoke. When your animals grow used to being handled and led they are ready to be put in a yoke. Keep one or both animals on a halter — in essence you will be leading the two animals in a yoke. Your goal is to get them to follow you willingly. Lead from a position near the nigh steer.

Step 4: Stopping and starting. Teach the team to start and stop on command. Use the lead rope as little as possible. The rope should not become a cue that the animals learn to depend on before obeying a command. Use it only as a safety measure to prevent your animals from bolting or running away. A team of oxen should never be allowed to run away. If they learn to run away, this dangerous habit is difficult to break.

Step 5: Turning. To turn your animals, think of each steer as a brake. Turn the team as you would turn a tractor or bulldozer by using the brakes. Slowing or stopping one animal will effectively move the team in the direction of the slower animal. If the animals are still using halters or lead ropes, make sure to keep them slack at all times. Most teams quickly learn to turn, if this concept of braking or slowing one animal is used.

Step 6: Pulling a load. When the animals are obeying all the commands for stopping, starting, and turning, it is time to teach them to pull a load or wagon. When hitching animals for the first time, be ready for them to try to run from whatever you hitch them to. Although cattle usually don't run far, they will run if given the chance. If you do not allow them to run away or shy from their first hitch, they rapidly become accustomed to pulling. Younger cattle quickly learn to pull a light load on a sled or cart.

Step 7: Practice often. Diversify your training or work routine. Your oxen should learn to depend on you for all cues. Do not allow them to respond to their natural instincts or their surroundings. Mix up the work, so that a strict pattern of daily work does not become what they learn.

Step 8: Avoid leading. As training progresses, avoid using or depending on halters. Tie the lead rope to the bows or yoke to keep from tugging on it to maneuver your animals. Start using no halter in areas where the animals are slightly confined, like a corral, pasture, or farm or woods road. Soon you will find that your animals can be controlled with no lead rope at all. Until you trust your team completely, keep the lead rope handy but avoid using it.

Step 9: Introduce challenges. When your animals become proficient with the commands and their wagon, sled, or implement, try more challenging commands like backing up, coming to you from a distance, or driving with only voice commands or movements of the whip. Although animals work best with both visual and physical cues, a patient teamster can achieve amazing results with minimal cues.

Cattle watch more than they listen, so be consistent in the way you talk, walk and hold the whip or goad when driving your team. Make sure your steers learn to respond only to your desired commands. They will try to anticipate your next move by watching for subtle cues. A pair of steers might start moving forward before you give the command, for example, because they saw you move forward. Make sure your team does only what you want them to. The same is true for moving around a barnyard or training area. Steers quickly learn where to turn to go home or to the place where they are unhitched.

Vary the workout so your cattle never know what to expect. Sometimes you will need to make steers do what they don't want to do. If they always try to run back to the corral where they are unhitched, make them back into the corral or walk in and out of the gate a few dozen times before unhitching them elsewhere. Each workout is a training session.

Be sure you are training your team and they are not training you.

Training oxen is an ongoing process throughout the animals' lives. A teamster who continually challenges the team and teaches them new commands will have animals that others envy.

Step 10: End on a positive note. Always end a training session after a success, and never after a failure. Cattle often condition their teamster to release them from work when they misbehave. The bad behavior is positively reinforced by the feed and freedom they receive upon their release.

Cattle readily learn to misbehave in order to avoid work.

Response in the Yoke

A common error in training cattle is expecting too much from the animals all at once. Cattle learn quickly, but only if they know exactly what you expect of them and are given consistent commands to accomplish the task or maneuver.

The training process must be sequential. Each part of their training should prepare the animals for the next step. Only after they can be directed to start and stop should you start teaching them to turn and back up. Once they understand and obey these basic commands you may add new commands.

During initial training your position as the teamster is critical. Although you can teach your animals to respond from many positions, most teams are trained with the teamster on the left side. This position allows the teamster good visual contact with the team, as well as the ability to hold the whip in the right hand to deliver physical cues.

Left-handed teamsters sometimes see this as a challenge. If the team is driven by others, however, or is later sold, most people will expect the animals to respond from the left.

Some teamsters work their oxen while standing in front of them, which can be helpful in slowing down a fast-moving team. However, driving from the front requires the teamster to turn around to see the animals, which can be tiring. In addition, the animals realize that the teamster has lost visual contact with them and will sometimes take advantage by following their own desired path of travel, which most often is toward home.

Teaching Commands

Commands must be used consistently. Teaching commands is a combination of good preparation, proper handling, and plenty of practice. No tricks are involved in getting your animals to learn and to

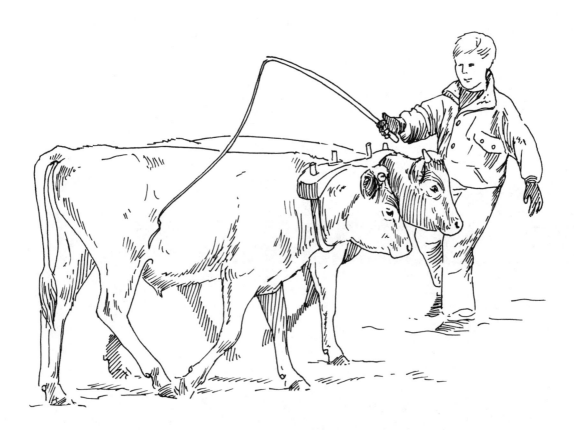

Leading from the front can be tiring because the teamster must turn to make sure the animals don't go their own way.

follow your verbal directions. Practice and patience will accomplish more than brute force and severe restraints. Learn to work with your animals and utilize their natural instincts.

Cattle can be trained the ways horses are, with lines and bits, and in many countries they are trained to work with nose rings. Such severe restraints are not necessary, but there are several reasons why a teamster might use bits, nose rings, and lines. Severe restraints are sometimes used by a teamster who has not spent enough time handling the oxen. In some countries this is the cultural tradition, where teamsters begin training the animals as mature cattle. The nose ring offers restraint and quickly establishes dominance. It may also be a failure to understand that alternative methods of training exist. Finally, maybe the teamster does not want to walk with the team and prefers to ride on the wagon or sled.

WHOA

The command to stop is by far the most important one you will teach your team. When you give this command, even by voice alone, your team should immediately stop all movement. They must learn to stop whether you are in front, back, or beside them. A team that fails to learn this command is dangerous in the yoke.

This command is important when capturing animals on pasture, haltering them, hitching and unhitching loads, and moving on roads, in the field, or on logging trails. This command must be taken seriously for your safety and that of your animals.

The whoa command should begin on the halter with each individual animal. Training should continue in the yoke. Before your team is hitched to their first cart or sled they should clearly understand this command. Be aware that the team may bolt when initially hitched to something they must pull. Prepare in advance for this possibility, by practicing moving forward only a few steps, and then using the command "whoa." Repeat as necessary until the team understands they are not to run away but must immediately stop, even when pulling.

Using a Goad

Each verbal command is accompanied by a corresponding way to move your body and hold or move the whip or goad stick. Your animals are thus cued by sight as well as by sound. The controlled use of a whip or goad will get the desired response from your team.

During initial training a solid stick or goad is better than a lash because it can more easily be controlled. A team soon wearies of a lash that constantly irritates them without giving them any clear cues as to what to do.

Use the whip or goad stick at first as a physical cue. Over time let it become primarily a visual cue.

To start your animals, tap them on the rump.

To stop, put the whip in front of them or tap them on the nose, head, or brisket.

To back, tap both animals on the knees or brisket.

To turn left, tap the off steer on the rump and the nigh steer on the nose, head, or brisket.

To turn right, tap the off steer on the nose and the nigh steer on the rump.

In time, a simple movement of the whip in the appropriate direction, short of touching the animals, will result in the desired reaction.

Movement of the goad stick in the appropriate direction will give you the desired reaction.

GET UP

The greatest variation among commands used by ox teamsters is in how they get their teams to move forward. Common terms include "step up" or "whaa-hoosh," making clicking or whistling sounds, as well as a number of other visual, physical, and verbal cues. Whatever cue you choose, use it consistently and do not confuse it with other commands.

When this command is given, usually one animal starts first. To accomplish heavy tasks like logging, plowing, or competition pulling, your team must be trained to move in unison and start together.

Like whoa, the command to move forward should work by voice alone. Only occasionally should you have to use the whip to remind an ox that is not paying attention. The animals often learn to watch you for cues to move forward. Be clear about this cue. For example, when training calves, teamsters often tap the animals with a whip to get them going. The cattle learn quickly that when the whip goes up in the air, it will soon come down on them if they do not move forward. The whip going up in the air becomes the cue to move forward as the animals progress in their training.

GEE

Pronounced like the letter G, this verbal command indicates a right turn. As you stand near the nigh or left ox, your team must move away from you. This command is generally difficult to teach animals that have become accustomed to following you on a halter. Wilder or less friendly cattle may see it as a chance to escape. Be prepared to control your team and do not allow them to take advantage of the situation during the turn.

If your team does not respond as desired in making a right turn, use the whip like a brake on the off (right) steer by tapping him on the nose, horn, or brisket. At the same time speed up the nigh steer with a tap on the rump, which will force the off steer around in a right turn. The verbal command should be followed with physical cues to get the desired reaction. This procedure may at first seem awkward, but will soon become second nature for both you and your team.

Practicing with a Pole or Chain

Once the team responds to the gee command, try it while they are pulling a cart or sled with a pole between them. Adding the load creates an additional challenge because the animals have the pole to contend with. Using a pole is preferred over using a chain, since green animals sometimes jump over a chain, get tangled in it, or turn around and face whatever they are pulling. In the gee turn, the off steer learns to step away from the pole. Tap the steer on the ribs nearest the pole and urge him to step out.

Training with a pole pays great dividends later when the team is hitched with a chain. Eventually the team should learn to make tight turns on a chain without stepping over the chain, better yet without bumping into it at all. A chain can cause serious injury to the hock and leg of an ox that is not taught to step away from it when pulling.

In showing oxen that are hitched with a chain, young teamsters are encouraged to completely stop the team before a turn. The team is directed to take a step back and turn to the right before tightening the chain. The teamster then encourages the nigh steer to stay close to the chain and the off steer to step away from it. Before giving the command "get up" the teamster positions the team at up to a 90-degree angle from the front of the sled or log. The off steer has stepped a few feet away from the chain. When the team moves forward the off steer has clearance from the chain as it moves toward him throughout the turn.

Another way to teach the right turn is to walk just ahead of the steers, again slowing the off steer with the whip and having the nigh steer follow you around to the right. This technique is usually successful with animals that have the tendency to try to run away to the right when directed to make the turn. The desired outcome in training the animals to turn right is to have the off steer slow down with the verbal command and the nigh steer pivot around him. Since you may encounter situations where you cannot walk ahead of your steers through a turn, they must learn to turn by voice and physical cues from the whip alone.

Some teamsters use variations to the gee command. Many say "gee off," since the team is moving away from the teamster. "Back gee" is a variation typically used for tight right turns. When this command is given, the off steer not only stops, but begins to back up, while the nigh steer turns sharply to the right. The off steer acts as a pivot and pulls the nigh steer around.

HAW

The haw command is given to make a team turn left. As you stand on the left side, your team moves toward you, following your lead to the left. When training young calves, placing the faster or stronger steer on the off side ensures easy haw turns.

Many young halter-trained cattle learn this command more easily than the gee command. As with the gee turn, use the braking technique if you encounter difficulties. Because the turn is to the left, the nigh steer is slowed or stopped and the off steer pivots around him. Use the whip if necessary to slow the nigh steer with a tap to the nose, horn, or brisket and tap the off steer on the rump to bring him around.

Variations of this command include "come haw" and "back haw." The latter is used for making tight left turns. The teamster backs the nigh steer as the off steer comes around in a tight circle.

This command and its cues become second nature to both team and teamster. The goal is a quick response from your team when you give just verbal cues and subtle visual cues. Over time your team will learn to anticipate a verbal cue from your body position and the placement of the whip in front of the nigh steer. In making the left turn, most teamsters move away from the team or toward the rump of the nigh steer to allow the animals room to turn left. This body language, although subtle, becomes an important cue to the animals.

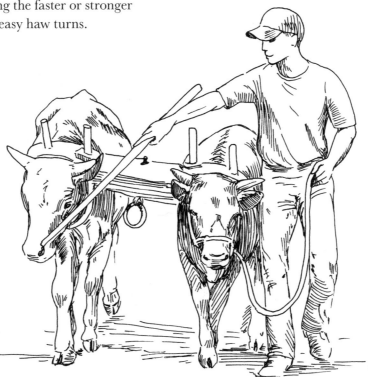

For a wide right turn, use the whip as an extension of your hand to slow down the off steer and speed up the nigh steer.

As with the gee turn, the haw turn should first be practiced without hitching the animals to anything. When they respond easily, continue training with the use of a pole on a cart or sled. In making the haw turn, be sure to step the nigh ox away from the pole, encouraging him to step out by tapping him on the ribs.

Later, when the chain is introduced, continue stepping the nigh animal away from the chain. The preferred method is to completely stop the animals. Make them take a step back until the chain is slack. Once it has some slack, turn to the left, with a tight haw turn. The team should be at up to a 90-degree angle to the load. Before tightening the chain and moving forward, make sure the nigh ox has plenty of room between him and the chain to avoid a leg injury. In New England competitions most judges discriminate against 4-H teamsters who allow their animals to touch the chain when making turns.

BACK

The back command is a challenging command to teach, so it is often neglected in early training. A team that does not learn to back up is almost impossible to hitch to wagons, sleds, or logs. Spending time teaching your team to back up is therefore as important as teaching them to turn right or left. Practice this command on a regular basis. An ox team very rapidly becomes reluctant to back up if this command is not a regular part of their training. Teaching your team to back up on a slope or hill is a great training aid.

As with other commands, the animals should respond to the back command by voice alone, but

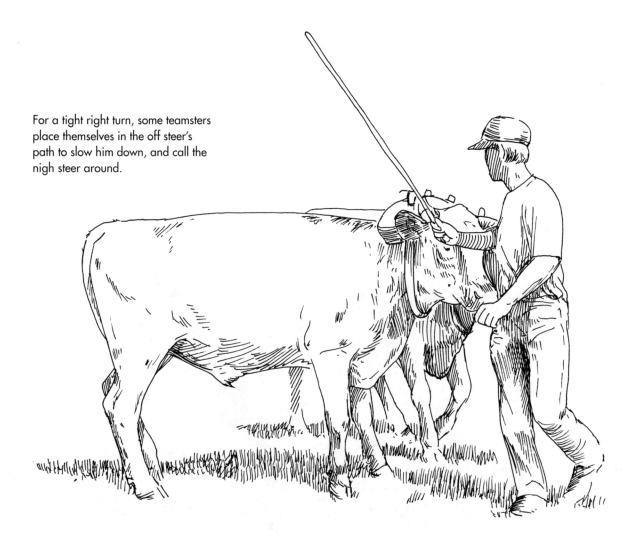

For a tight right turn, some teamsters place themselves in the off steer's path to slow him down, and call the nigh steer around.

physical and visual cues are important during training. Initially tapping the animals on the knees or brisket will cause them to step back. Try not to hit the animals in the face, which will encourage them to put their heads down and refuse to move. Standing in front of a young team and facing the animals while saying "back" intimidates the team into stepping backward. When they begin to understand what you desire of them, return to the left side of the team.

Your team must be taught to back up straight with their bodies together. Their initial reaction will likely be to separate their rear ends and try to turn in the yoke. Use the help of an assistant or nearby fence. If the cattle are tame you can tie their tails together with light string or the hair of their tails. Do not use heavy twine or rope, and do not tie the tails of flighty animals, as they may injure or break their tails.

Practice backing in a straight line. As a guide for training the team to move straight back, mark a line on the ground, lay down planks, or follow the edge of a road or path. The oxen should learn to back to a pole of a wagon or cart, as well as backing to a log or sled where they will be hitched with a chain.

A well-trained team can back up when you stand behind them holding the chain. Try not to pull on the chain, which will become a physical cue the animals learn to expect. A reasonable goal is to have a team that easily backs up in a straight line for 100 to 200 feet. While such a goal may at first seem daunting, a team that can back up this well will never have a problem hitching to anything where backing is required.

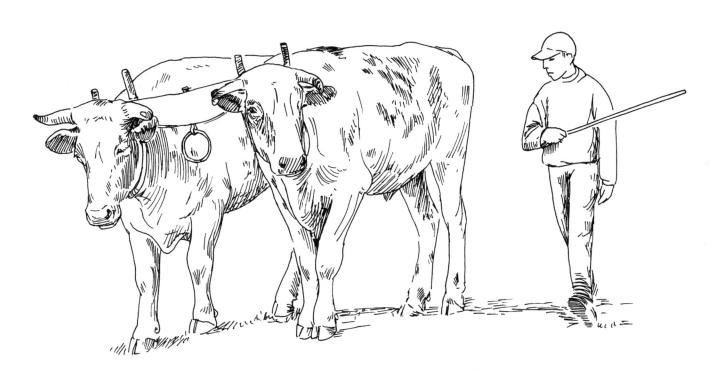

For a haw turn, the teamster steps back and away from the oxen to allow them room to turn left; the nigh animal is slowed down and the off steer is called or sped up with a tap of the whip or goad.

Backing with a Cart

As your animals learn to understand how to back, begin training them to back up while hitched to a light cart. As in other phases of training, they will initially balk at this request. Using a hill or slope is again helpful. As the team becomes older and develops horns they can be taught to back up large carts using only their heads and horns. If they are to back a really large wagon, you may choose to use breeching like that used on a horse.

STAND

When you give the whoa command you should expect your team to remain in place and attentive until you give another command. To the novice teamster this goal may seem impossible. Similar to having your team back up 100 feet with voice commands, having them stand is not only possible but is an important step in proper training.

Teach your team to stand still a little at a time. Use the command "stand" or "stand still." Every training session or workout should include some short breaks when your animals are expected to stand in the yoke.

As your team learns to stand still, start moving away from them. At first move around the team. As they become more trustworthy move farther away. If the team seems to accept standing alone, try stepping out of sight behind a wall or barn door where you can keep an eye on your animals. If they move, step out and correct them, either with a verbal command or a physical cue. Before trying this, make sure your team can be trusted to stand. Otherwise you might give them a chance to learn they can run away.

Oxen can be trained to stand still even when they cannot see you. But leaving your team unattended is not a good idea in public or if the team is hitched to a valuable implement. At times you may have to go out of your team's sight to get a tool or make adjustments to an implement. Having your team remain in place can be helpful around the farm or forest. In the show ring a team that remains perfectly still and then comes when called impresses the judge and audience.

OTHER COMMANDS

Whoa, get up, gee, haw, back, and stand are the six most common commands used for working cattle in the United States. The six basic commands are necessary for a team to function on the farm and in the forest. They are relatively easy to teach. Before a team is used for real work, they must understand these six commands. But these six are only the basics. The number of commands an ox can learn is limited only by human patience and time.

Teamsters around the world use other commands in their native tongues to get the same reactions from their animals. The word used is not what really matters. More important is making sure the animals understand the cues. Some teamsters make up their own terminology and visual cues. Nothing is wrong with making up your own verbal cues, provided your team will not be sold to someone who understands the traditional language of training and working cattle.

CHAPTER ROUND-UP

Having a logical plan for dominating and teaching your young oxen takes the guesswork out of everyday workouts. Before long your team will behave in a way you never thought possible. Most halter-trained steers make tremendous progress in just one month when worked for an hour or two each day.

A well-trained team is the result of hard work, many hours, and a firm but patient trainer. Challenge your steers often, in both their training and their conditioning. Ox training never ends. Asking your animals to do new tasks is good for a team of any age. A pair of steers can learn a lot in a month, but you will be amazed at what they will learn in five or ten years. ★

Bud Kluchnik
Ripley, Maine

I first met Bud Kluchnik at Tillers International in the late 1990s at a Midwest Ox Drover's gathering in Kalamazoo, Michigan. In his thick Maine accent Bud introduced himself as the guy who owned the head-yoke oxen I showed in my first book. We had a great week together, and I learned a lot from Bud about shoeing oxen, training steers, and making head yokes.

When I stop to see Bud in rural Ripley, Maine, he always greets me with a warm smile and a trip to the ox barn or pasture. My most memorable visit was with a friend who had grown up in the suburbs in the Midwest. He was new to New England, and I knew he would enjoy Bud. His comment at the end of our visit was, "This guy is the real deal."

Asked how he got started with oxen, Bud replied, "I remember going to the Fryeburg Fair when I was eight or nine years old. My mother gave me $10 for the day, but I hardly spent any of it, because I just sat outside the ox-pulling ring and watched oxen all day long."

He really got started with oxen when he was about twelve years old, living in Windham, Maine. "There were two or three old fellas around who had oxen," Bud said. "One was John Richards, and he and his son had head yoke oxen."

As a boy, Bud used to go to all the fairs in southern Maine. Sitting on the fence at Cumberland Fair one year, he watched the Nova Scotia oxmen with their teams, down for the annual International pull.

"Those fellas had apple-fat oxen," he recalls, "all shod up, and dressed up with those head yokes and pads, and they pulled without a lot of fuss. Right there and then I knew that I was going to have a team of head-yoke cattle."

Bud went home and took a single Jersey-cross steer that he had and proceeded to make a head yoke. He said, "I took a 4 × 4 piece of hardwood and carved a yoke. I didn't know what I was doing, and I was lucky I didn't pull his horns off. The yoke looked like one of those crude Mexican yokes, and I hitched him up. He would pull — wouldn't he pull."

Bud tried pulling oxen competitively, but never really put in the time he needed. He has enjoyed going to fairs showing his oxen and helping others get started. "I enjoyed the kids the most," he says. "Chad, Spencer, Seth, Sean, Myles, Matt: there were a lot of them. Kids have a mind like a sponge. It's easier for them to grasp new information, but as we get older, our mind is more like a stone, pretty tough to take in anything new."

As far as training, Bud admits, "I do most of my training in the barn. Winters are long up here, it seems like winter is all we have sometimes. I like the calves to be hooked up in the barn. I want them to be handled. I back them up, move them over, and just work around them. I really just fool with them, and then teach them to lead, tying them together with a piece or rope or twine by their collars. They have to follow you and listen to you, before you yoke them up."

Bud has used a bow yoke but truly prefers the head yoke. He states, "I really don't even start training them in the yoke as calves. I can make a head yoke for a team of calves with horns 3 inches long, but many times I do not yoke them up until they are a year and a half old or even two."

Bud offers the following advice for anyone starting out. "Just ask them to do what you want, and then give them a little whack if they need it. The calves catch on fast. Don't get discouraged."

6
ADVANCED TRAINING

Most teamsters lack the commitment or the reason to train their team to function at more than the most basic level, but oxen may be trained to do all sorts of tricks. Bowing, jumping over things, lying down, playing dead, and similar tricks are usually performed for entertainment. Side-stepping or backing from behind without reins, on the other hand, are tricks that can help you direct your team while working. If your oxen are going to work, they must be trained to pull wheeled vehicles or farm implements and dead weights such as logs, sleds, and plows.

Pulling

After a team are trained to respond in the yoke, or are "handy" as we say in New England, they may be trained to pull. This training may begin at any age. Teams a few weeks old can learn to pull a pole or a tiny sled. As you would in your own exercise program, begin with light work and gradually increase the load. Your animals need time to become hardened to the work and accustomed to the yoke.

Heavy pulling too early or with poorly fitting equipment creates animals with sore necks, discouraged attitudes and an unwillingness to pull. Just like a weight lifter who overdoes it on an initial workout, a young ox may run and jump easily when first put on a load, but his muscles will soon fatigue and become sore. Steers quickly become discouraged if you start them with loads that are too heavy for their undeveloped muscles and bones.

The initial load should be light and easy to pull. Five to 10 percent of the animals' combined bodyweight is usually a safe amount of weight to begin with. As the animals become acclimated to pulling, gradually add more weight over a period of weeks.

Training a calf to pull is more an exercise than real work; the calf should be asked to pull no more than he can easily move, or about 10 percent of his bodyweight.

Mature animals can physically tolerate a higher percentage of their body weight, but should still initially begin with no more than 10 percent of their combined weight.

Young growing steers take longer to learn to pull because the amount they can pull is limited by their body size. Frequent workouts pulling a light load and plenty of rests go far in avoiding discouragement in a young team. Light loads condition the animals for heavier work. The young team needs to build skills and stamina and become accustomed to wearing and using the yoke.

PULLING EQUIPMENT

You do not need an elaborate sled or wagon to teach a team to pull. One of the most common devices for a young pair of calves is an old car or truck tire without a rim. Such tires pull easily, wear out slowly and, as the animals grow, may be increased in weight by adding rocks or concrete to the inside. As the team grows, tractor or loader tires would provide extra weight for a good workout. Many New England ox teamsters have a collection of sleds, stoneboats, and old tires that allow them to choose a load of the right size and weight for the team being worked.

When training a young or unruly team, train them using a pole instead of a chain. The pole is rigid and will help teach the animals to stay on their respective sides when the chain is introduced later. The pole is also less abrasive on the animals' legs as they learn to turn corners with the load. A small hardwood pole that can be attached directly into the yoke ring, with a hook or short chain at the other end to hitch to the load, works well for training. Although many teamsters are successful starting with a chain, for the novice the use of a pole eliminates initial challenges such as having animals jump over or tangle in the chain.

Give your animals adequate room between themselves and whatever they are pulling. A short chain or pole allows the animals to get more lift on the front of the load and reduce friction, thus making the load easier to move. A young team, however, needs space to allow for a lagging animal and for turns. Animals that catch their hooves under the sled or log will be injured. Calves need a cou-

Stoneboats

Flat wooden stoneboats were once common on New England farms for removing stones from recently cleared or plowed fields. They are still used today, but are often made of steel or wooden planks with a curved steel head. A stoneboat gives the team a good workout while allowing the teamster to easily load and unload heavy stones and concrete blocks.

ple of feet between their hooves and the load. The required space for older cattle may be 3 to 3½ feet, as their legs stretch out more when pulling. As the animals mature and grow accustomed to pulling, the chain may be shortened. Monitor your chain length at all times.

Never start a team pulling with a slack chain, which may injure the animals and/or break the yoke or bows. Steers often sense slack in the chain and do not expect a load. As they walk forward they may hit the load and possibly bruise a shoulder, stop due to the shock, or lunge ahead and create a tremendous momentum that leads to a breakdown. Oxen are patient animals. They will learn to lean into a load and then push forward and upward. This slow but deliberate movement allows them to pull tremendous amounts of weight without injury or breakdowns.

PHYSICAL CONDITIONING

Oxen are draft animals trained to do work. Learning to pull is the basis of their existence. An ox is like an athlete in many ways. Pulling logs, plowing, spreading manure, and moving stones piled high on a stoneboat are difficult jobs. The animals need to be physically and psychologically ready for such tasks.

Much like a human working out at the gym, physically conditioning oxen for work requires time and regular workouts. Challenging the animals to pull tremendous weights for short distances is not desirable in early training. Even mature oxen quickly become soured on pulling if expected to pull heavy loads all the time.

A team that is worked irregularly or seasonally requires a few weeks of conditioning before they can put in their best performance. A team that is rushed into a task without adequate training or conditioning will resist the work, become reluctant to pull, and cause a great deal of frustration. Preparation for the task at hand through conditioning is critical. Conditioning takes at least three weeks before a team is ready to do heavy farm work, longer before the team competes in pulling competitions.

When teaching oxen to pull, realize they cannot tolerate as much heat as horses or mules can. Many oxen have died from heat stroke. Cattle do not sweat like horses and cannot cool themselves as effectively. Cattle that pant in hot weather are trying to dissipate extra body heat. While a little panting is not critical, it is a sign that the animals need to slow down, rest, drink some water, or be cooled off by some other means, such as being sprayed with water. Cattle that are accustomed to working in hot weather proceed at a slow pace, limiting their exertion to prevent getting overheated.

Bodyweight and Work

Conditioning includes not only proper work and training, but also getting the animals physically ready to work. Animals that are extremely thin or overweight will have to be fed accordingly to maximize their performance. In many developing countries cattle are thin at the beginning of the rainy season when plowing begins. Similarly, in the United States cattle were historically thin at the beginning of the plowing season, after a long winter on poor feed. Feeding the team so they gain weight and stamina is an important part of conditioning. Emaciated animals cannot be expected to exert their full potential.

Overweight cattle, on the other hand, perform best after losing some of their excess fat. An overweight team must be trimmed down if they will be worked in hot weather. Cattle carrying a few hundred extra pounds suffer from physical exhaustion, as well as heat exhaustion, more readily than thin cattle. Many teamsters prefer to keep their cattle thin, then feed them to gain weight during conditioning and heavy work.

Oxen that will be used for heavy farm work or competition pulling need regular exercise and conditioning.

Training a team to pull heavy loads is one of the most challenging parts of training for a novice teamster. Oxen would rather sit in the shade and chew their cuds than work. They get especially lazy in hot humid weather or when the work is difficult. Each team has to be taught to really dig in and pull when it is necessary.

One trick is never to let the team realize their limits. Sometimes the teamster sets the animals' limits without knowing the team's true ability. If a team finds out they can bluff the teamster into believing they are too weak to move a log or pull a plow, they will do it. Sometimes they have to be forced to realize their ability. There is a fine line, however, between a team that is ready to be challenged and a team that is not in good enough physical condition to attempt heavy loads.

LOAD GUIDELINES

Teaching your team to pull, or building the stamina of a team that already knows how to pull, takes time. Repetition is the key to getting the animals to perform well in the yoke. Use light weights to accustom and condition the animals to the task while they build strength and endurance. Gradually work the animals up to heavy loads. A team should pull heavy loads only after they are proficient with light loads. Plowing and logging are among the most difficult jobs an ox performs, and conditioning is essential before these tasks are attempted.

The basic guideline used in New England for short hauls, with regular rests, is that a well-conditioned team should be able to pull half their weight while logging in the forest or pulling a heavy sled. Provided the footing is good and the path is clear, the animals will catch their breath on the return trip into the forest while not pulling a load.

For competitions, expect mature oxen to pull one to two times their weight in stone or concrete on a stoneboat or sled for short distances, such as a few hundred feet in a few minutes. Such physical performances have been going on for hundreds of years in New England. The challenge of competition has raised the level of expectation beyond believable weights. This level of expectation is usually more than the average farm or woods team can sustain.

It takes incredible training, conditioning, and culling animals without the will to dig in, in order to get two oxen to achieve such feats.

PULLING TIPS

Pulling incorporates many training components that go beyond basic training, most of which come only with experience. Cattle may be used for pulling competitions, show and farm, or forest work. If you are interested in pulling competitions, realize that many secrets have been developed to get cattle to give everything they can when pulling. Many ox teamsters who compete regularly cull out animals that do not have the psychological will to put all their strength into the yoke.

Many champion pulling teams are specialists. They have been specifically selected, trained, and conditioned to pull tremendous weights. Even so, the novice teamster should not be discouraged from competing. You can learn a lot from watching or competing in pulling events, and few teamsters step into the pulling ring and win a blue ribbon with their first team.

The following pulling tricks may save time and effort for both you and your team:

Cutting the Load

Getting a load moving takes more energy than keeping it in motion. A heavy log, sled, or stoneboat starts easier when the load is "cut," or pulled at an angle. Unlike starting straight forward, cutting the load requires setting the team up. For cutting to the left the nigh steer needs to be stepped away from the chain. For cutting to the right the off steer needs to be away from the chain. When the team pushes forward the chain and load straighten out and the animals' legs are not in danger of injury. If a log, stoneboat, or sled is frozen to the ground, cutting the load at the start of the pull makes a tremendous difference.

Making Sharp Turns

When making sharp turns with a cart or stoneboat using a pole, keep the animals' legs away from the pole. Animals that lean on the pole or wait to be pushed by it hinder quick, sharp turns. Although

Rules for Teaching Cattle to Pull

ALWAYS:

1. Start teaching a team to pull using a light load.

2. Make sure the team is able to start the load. If the team fails to pull the load on the first try, lighten the load or check for something that might be preventing it from moving. Logs with branches stuck in the ground, sleds frozen to the ground, or rocks jammed against a wagon, log, or sled: these are commonly blamed on a team thought not to be trying hard enough to move the load.

3. Have the team pull short distances with frequent rests. Rests keep the team from being overworked, and frequent starts and stops are a great way to teach them to start the load together.

4. Begin training with short daily or twice-daily workouts. As the team becomes acclimated to the work, gradually increase the amount of time they are expected to pull. Keep in mind that calves less than two years old do not have the physical or psychological stamina of older cattle.

5. Make sure the yoke and bows properly fit the team; constantly monitor their comfort and fit.

6. Make sure the chain length is appropriate for the team's size and load.

7. Keep the steers' legs away from the chain, especially when turning.

8. Hitch to the load quickly and easily. Getting the team excited before hitching may get a little more initial energy out of them, but they will quickly tire of the game.

9. Start the load on an angle by cutting or freeing the load, rather than pulling it straight forward.

10. Make both teammates pull together.

NEVER:

1. Overload a young or inexperienced team.

2. Attempt to work a team where they could be injured.

3. Overuse the whip to encourage a young team to pull.

4. Pull long distances without frequent rests.

5. Work a young team too long for their abilities.

6. Use a yoke or bows that do not fit the team.

7. Turn corners too tightly, causing the chain to rub on the animals' legs or catch a foot under the load.

8. Allow a team to stop unless they are directed to do so.

9. Work a young team in deep snow or mud, or on ice.

10. Work a team when you are in a bad mood.

---

the pole is less likely to cause injury than a chain, animals that lean on the pole may be tripped on a sharp turn.

A cart or stoneboat may be turned 360 degrees simply by pivoting the team. A cart may be turned around by pivoting one wheel without moving forward or back. When the team is well trained, the pole never touches the animals' legs.

Whether the pole touches the oxen is often judged in ox shows to demonstrate how well the team can turn a cart. A frequently added dimension is to put a paper plate under the wheel that is pivoting and never have that wheel come off the plate during the 360-degree turn. A similar turn may be made with a sled or stoneboat without having the animals' legs touch the chain.

CHAIN LENGTH

When hitching a team for pulling allow enough room between the steers' rear feet and the front of the sled or drag to prevent them from catching their heels during the pull, especially when turning. One common method used by youngsters to check the chain length is to pull the chain straight back from the ring where it is hitched to the yoke and wrap it around a steer's thigh until it touches the flank. This chain length will provide plenty of clearance for the sled to be far from the animals' hooves when turning. However, for heavy pulling, keeping the chain shorter than this will make pulling the load less difficult.

Advanced Tricks

Ox competitions have created a challenging environment where teamsters strive to train their oxen to do something no one has ever seen. Oxen have been trained to step sideways, respond to all their commands without wearing a yoke, be ridden, and use their heads to lift their teamster. Frank Scruton of Rochester, New Hampshire, once taught two oxen to teeter-totter while either standing or sitting on a wooden plank.

Given lots of time and patience, oxen may be trained to do almost anything other animals are able to do.

As with all training, time and patience are the keys to teaching your team to do tricks. A team generally works best if taught one new command or trick at a time until that trick is mastered. Tricks, if not practiced until they are fully mastered, are quickly forgotten.

STEP IN AND STEP OUT

Most oxen easily learn to step in and step out, which means the ox steps to the right or left with its back feet. Begin training by getting the animals to step in toward one another or step away from one another when in the yoke or tied in the barn. Most teamsters use this maneuver when grooming or examining their animals.

Stepping out comes in handy when hitching the team, to allow easy access all the way to the yoke without getting squeezed between the animals. You can also get an ox to step away from, or step toward, the chain or pole during a turn. The teamster typically calls the animal's name and then directs him to step in or out.

SIDE-STEPPING

Teaching your team to step sideways is useful when hitching to a log or cart. Side-stepping allows for accurate positioning when backing a cart or wagon and may be used to position animals for photographs or to move them away from vehicles, buildings, or other objects. Side-step training begins after the animals have mastered stepping in and out. The animals are encouraged to step to the right or left using their front feet as well as their back feet. This maneuver may at first seem daunting and does require a great deal of patience. Once the animals understand your intention, however, they will side-step easily.

BEING RIDDEN

In the United States, oxen are usually ridden solely to show off the team's abilities, but cattle are ridden in some countries as a means of transportation. While their gait is not as fast or as stylish as a horse's, and horse saddles are not designed to fit them, cattle can readily be trained for riding. They most easily respond in the yoke while directed by the teamster using a whip and voice commands. They may also be ridden singly and directed with reins attached to a halter, hackamore bridle, or nose ring. I don't recommend a nose ring because it is a severe restraint that usually causes the animal to become sore and shy around the nose.

DRIVING FROM BEHIND

A team that is trained to drive from the side and respond to the basic commands may easily be taught to drive from behind. Small calves may be taught to drive from behind as easily as from the front or side without reins or ropes. Older animals require more restraint because they are more eager to run away. Begin this training in a confined area and get the team sufficiently tired beforehand so they will not be inclined to run away.

Initially the team may confuse your request to move ahead with your movement to the back of the animals. Work the animals on a trail, path, or farm road heading toward home. Most teams are eager to return home and will readily trot along. Once they get the idea, try them going away from home or in the field following a furrow or a freshly cut swathe of hay.

Driving your team from behind comes in handy when you are working in a woodlot, mowing hay, plowing, or using other farm machinery. Without reins you have to rely on voice commands and/or a long whip. Training to drive without reins should not be done in an area where the animals might possibly run away, destroy property, or injure people.

In other parts of the world many teams are trained to drive from behind, guided with reins attached to a halter, to a bridle with a bit, to the nose using a nose ring or rope, or even the ears. A halter is the least severe form of physical restraint; a bridle with a bit may create problems with cud chewing. I do not recommend using the nose or ears, commonly seen in Africa, because doing so makes the animal head shy.

USING NO PHYSICAL RESTRAINT

If you can direct your team without a yoke, halters, lead ropes, or other physical restraint you have achieved a high level of training and trust. Animals

Once calves have learned to respond in the yoke and pull a cart and drag, they may be taught to be driven from behind; make sure they understand the basic commands from the ground before attempting to ride.

that follow willingly or obey directions without any being physically forced to do so have learned to trust and respect the teamster. The teamster in turn has established a unique relationship with the oxen. Having a team that willingly follows is helpful in catching animals in pasture and walking the team to water. 4-H teamsters walk teams without restraints to demonstrate their ox training abilities.

CHAPTER ROUND-UP

The patience and firmness you developed while teaching your steers the basics have carried you through this more advanced training. Having come this far, you no doubt have convinced your oxen to trust you and look to you for commands and guidance. In turn, you are now able to understand how your steers think and react to new training principles. Once your oxen are trained to function in the yoke and respond to human direction, the only limits to the number of things you might teach them are your ability as a teamster and the amount of time you are willing to put into their training. ★

Frank Scruton

Rochester, New Hampshire

It was the height of the Great Depression. Frank Scruton remembers people coming around the farm looking for work in exchange for food. During those difficult winters his dad used up to six oxen to haul firewood by sled, two cords at a time, to the townspeople of Rochester. It was an all-day trip, more than 10 miles each way. He says his dad, Arthur G. Scruton, would leave before dawn and would return after dark, day after day. "It was like that back then, tough times, a man had to work hard if he wanted to make do."

Frank learned ox driving from his father, when just a boy. When asked if Arthur ever gave him advice on driving oxen, Frank answers in his squeaky voice, "He gave me pointers all right, all the time. He really knew oxen; he'd been driving them all his life. He often kept me out of school to drive four oxen in spring while he ran the plow. But that was the way in them days."

He can't remember when he drove his first team, and when asked how many he has trained over the years, he says, "I can't remember, but I have had a lot of them." As you look around his house, perched on a hill overlooking the farm, it is obvi-ous he has had a lot of oxen. There are photo albums, pictures on the walls, and other oxen memorabilia all over the house. Outside, an ox yoke hangs over the garage, and his teams are across the street.

Most impressive are two quilts Pauline made from the ribbons Frank has won. In 1984 he was undefeated with Babe and Tom, a pair of Chianinas that are obviously among his favorites. Hanging on the wall is a large quilt made of blue ribbons won that year. "Frank likes the blue ones," Pauline says, "but I prefer the multicolored ribbons."

Into the Limelight

Frank became famous for his oxen during the Depression. As he tells the story: "I was doing tricks with a pair of Milking Shorthorns [he calls them Durhams] named Line and Swan at the Rochester Fair. They were a pair I trained in a sawdust shed over the winter. [Frank had appendicitis and so he wasn't allowed to work, but training oxen is fun for Frank, not work.] They were just calves, but they could do all sorts of tricks. Then this fella comes down from the grandstands and wants to hire me for the Vaude-ville Acts for $200 per week. That was a lot of money back then."

The next thing Frank knew he was traveling the country from July to October, displaying his talents as an ox teamster for tremendous crowds, often sandwiched between elephant and bear acts. Frank did this for five years. He said "If the war [World War II] hadn't come up, my act was about to be expanded, from steers doing tricks on a teeter-totter and playing dead to a steer diving off a high board."

He trained them to lie down, sit, climb on a teeter-totter — they could rock back and forth with their front feet only, their back legs only, or all fours. They would play dead, and, most unique of all, Swan would jump over Line on command.

Frank's vaudeville career might have been halted before his dream was fulfilled, but he seems to have gone on to fulfill a lot of other dreams. When asked if he offers any advice for teamsters just start-ing out, Frank offers only this: "You have to be patient and put the time in on them. My father always said be sure they are trained before you go to the fair."

Frank said it wasn't too hard to train Line and Swan to do all those tricks. "I just bent their legs and forced them to lie down when they were calves, all the time saying 'lay down.' You just keep doing it until you get it in their thick heads."

Star and Bill, a pair of Holsteins, and Babe and Tom, Chianinas, were his favorites over the years. "Chiani-nas are the breed to go with these days; that's where the money is. Chianinas are top, no doubt about it. Other breeds like the Holstein or the Brown Swiss don't have the zip or the size of the Chianina. There's nothing wrong with a good Holstein or a Shorthorn, they will work all right — although Brown Swiss can be a little lazy — but I have had some good ones of about any breed."

A Changing Scene

Ox pulling has changed, according to Frank. "Years ago you would walk your steers to the fair. There wasn't the competition you see today. It's nothing for an ox teamster to take a load of steers to a pull 200 miles away. The trucks are bigger, the roads are better, and I do it as much as the next guy.

It wasn't like that in the old days. You pulled against your neighbor, with cattle you worked all year. We hauled out wood all winter and worked cattle all summer. Until trac-tors came around after the war, we used oxen for everything. These days I just work cattle in the summer to get them ready for pulling in the fall. I don't have the energy I used to. I probably ought to have just one team and focus on them, but I have three or four pair instead."

As I drove home, I remembered when I first saw Frank Scruton com-pete with his oxen. It was the New Boston Fair, in a little pulling ring tucked in the woods. He intrigued me 35 years ago with the ease with which he handled his team. Equally impressive was the tremendous load that they drew for him. Frank was always the gentleman and always a great teamster. I left now know-ing his style in the showring was perfected a long time ago, when he entertained much larger groups than the one at my home county fair.

7

TRAINING
MATURE CATTLE

Not all teamsters have the time, money, or desire to start training oxen as calves. Although cattle are easier to train as calves, adult cattle also have the capacity to learn to work in a yoke. Starting an animal's training when he is closer to maturity requires a more advanced level of skills and techniques than would be needed by a teamster working with calves.

Waiting until cattle are mature to begin their training has both advantages and disadvantages.

Some of the advantages include:

★ The return on your invested time and effort may be more rapid, *assuming* the training is successful.

★ Older animals are less likely to become weary and need frequent rests.

★ You won't need the numerous yokes of different sizes that are required throughout the training and development of young animals.

★ Older cattle may be more easily maintained on a diet consisting primarily of roughage.

★ The team is easier to match based on temperament, conformation, and size.

Disadvantages include:

★ Older animals are more expensive to acquire.

★ Your health and safety are at greater risk when you work with mature wild cattle.

★ You must have equipment to accommodate the uncontrolled force the larger animals may exert during training.

★ Older cattle may need a longer period of training before they can be trusted in the yoke.

★ You must have greater skills and more experience to train and handle older cattle.

Initial Acclimation

Cattle with long flight distances, showing an obvious fear of humans, may need a period of acclimation and mild restraint before being handled. This period varies from animal to animal, but may be days or weeks. Your initial objective is four-fold:

1. to calm the animals;

2. to reduce their flight distances;

3. to make them realize that they need not fear humans;

4. to let them become familiar with you as their trainer.

Place the animals in a corral or barn where all their feed and water are provided. Make sure they cannot escape, which would allow them to realize some dominance and independence. If you are training a single animal, pen him alone; if you are training a team, pen the teammates together. Do not corral the animals with a larger group, or they may not associate their captivity with humans.

Make your initial contact with them as pleasant as possible. Do not beat or push aggressive cattle into submission; doing so only reinforces their fear of humans. Being enclosed is stressful to animals that are used to freedom. Do not continue training until they seem at ease with their surroundings.

When properly trained and matched with the right teammate, a wild and aggressive steer may turn out to be the best animal in the yoke. Beware of animals that represent either extreme in temperament. Cattle that are extremely aggressive or extremely shy will never make good work animals.

Capture and Restraint

Once the cattle have been acclimated to being near people, use adequate equipment and appropriate methods to capture and restrain them. Time spent preparing a corral, ropes, and hitching posts to withstand the most unruly animal will pay great dividends. The worst thing that could happen during an initial capture is to have an animal break loose or realize his superior strength. Cattle that fear humans quickly learn how to escape.

The best method of restraint is one that is fast, effective, and safe. Capture in the United States usually involves the use of a corral, a chute, and a locking headgate. The animals are caught in the corral, moved through the chute and individually captured in the headgate. They may then be restrained by halters, ropes, or nose rings and tied to be allowed to fight their restraint. Cattle that have been properly acclimated to people and their new surroundings will not fight for long.

Never leave a green animal tied and unattended. He may choke, suffocate, or get tangled in the ropes. The most effective way to train cattle is to keep them restrained in close quarters after the initial capture. Cattle in the northeastern United States are usually kept in a barn until they can be handled and touched, and no longer resist being tied.

Initial Training

Many teamsters believe that it's best to start animals individually on a halter before yoking them as a team. Halter training allows the teamster to work with and evaluate each animal and allows each animal to become accustomed to being handled.

Some teamsters, on the other hand, feel that a more effective technique is to work the animals in a training ring without a halter before yoking them for the first time. Training in a ring has the distinct disadvantage that the oxen learn to associate only the ring with being in the yoke and responding to the teamster. Their first few sessions outside the ring will be challenging and will require a lead rope.

Still other teamsters begin training by tying a green animal to a trained animal. In Africa, and historically in parts of the United States, some ox teamsters turn the tied-together animals loose to learn to get along and walk together.

However, real training begins after the animals realize that they have to respond to humans. This often begins by learning to respect a tug on the rope they are tied with. Each animal as an individual must learn to respect a halter and lead rope.

Cattle weighing in excess of 3,000 pounds may be haltered, led, and handled only because they believe humans to be dominant; if this ox just once learns he can break the halter, no normal halter or rope will hold him in the future.

A **sturdy rope halter,** which applies pressure to the nose when drawn tight and releases when the animal gives in, is strong enough to hold an older calf or young steer.

A **heavy multilayered leather halter** has chains that draw tight if the animal pulls or resists. Such a halter is appropriate for a larger animal that has learned to break a less sturdy halter.

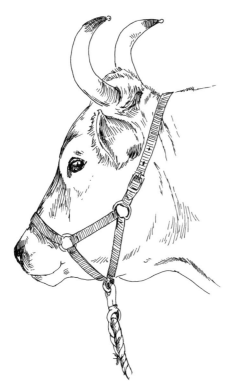

A **stitched nylon halter** offers a rugged alternative to leather, but is not suitable for halter-training older animals if it lacks the option of tightening on the muzzle when the animal does not respond.

Hitching the ox by his horns saves money, as no halter is needed, but it gives the animal so much power that the trainer will have no leverage in holding him.

GET UP AND WHOA

Initial training on a halter should include the commands to start and to stop. Make sure your physical, verbal, and visual cues are all consistent and easily understood. When you say "get up," say it the same way each time and follow with the whip going up in the air. Touch the animals on their rumps to get them to go forward.

Most teamsters stand on the left side of their animals rather than in front of them. This position aids in getting the team to move forward. Standing to the left side may become a cue for them to move forward or pay attention, but they should not move forward until they are commanded to do so.

To teach the command to stop or whoa, say the command and follow it by bringing the whip down in front of the animals and touching them on the noses or knees. You have given verbal, visual, and physical cues. Some teamsters also stop walking; the team quickly learns to stop when the teamster stops. Teaching your team to expect this additional cue may be undesirable if you hope to eventually drive from behind.

Too much whip in the face or on the nose early in training causes cattle to toss their heads and become head-shy.

When the team is first taken out of the corral, they should continue wearing halters until they learn to respond in the yoke.

Don Collins, a veterinarian and accomplished teamster from Berwick, Maine, has stated:

"Training a team of calves involves first teaching the animals to function in the yoke, and later teaching them to work on the cart or drag. But starting with an older team, it may be easier to hitch them to a load, teaching them to pull at the same time you're teaching them to work in the yoke. [The weight pulled by the team becomes a system of restraint.]"

Exercising Dominance

Animals that have been captured, restrained, and not allowed to escape during halter training have learned that you can dominate them. While this method of teaching dominance may not be the swiftest, it rarely causes animals to lose their spirit. In other parts of the world harsh training often results in animals that learn to lie down in the yoke.

Some methods of exercising dominance are so severe that the team becomes lethargic when placed in the yoke. Lying down on the job is not the desired result of dominance training. Harsh methods of restraint — such as the "running W" *(see* Glossary*)*, roping a foot, or beating an animal severely — often have the opposite of the desired effect. Instead of motivating the animal, these techniques teach him to lie down in order to resist being trained.

TRAINING METHODS

A training system that encourages rather than discourages the animals is to use an experienced team ahead of green animals, with a light load behind them and a pole between. If the animals are too wild or unruly to hitch, or no other team is available, physical restraint such as the running W or other rope system may be warranted, but these methods are extremely severe. I prefer to hitch the animals and allow them to pull a load to tire themselves out. Once tired, they are more likely to pay attention.

Some teamsters hitch the green team to a load, in a fenced corral or pasture, and fight the steers until they obey the commands to start and stop. This system is referred to as the "cowboy style." Even if you are using the cowboy style, have a goal and work toward it. Be sure you are not just chasing the steers and confusing them. If the animals do not start and

A green pair of young Dutch Belted steers hitched between a larger team of Holsteins and a sled, the sled's pole serving as a training aid. This system initially keeps the green animals under control, but they must eventually learn to work as a pair, without another team ahead of them.

stop on command after the first few sessions, you may need another method or some helpers.

Another option is to yoke a green animal with one that is already trained. The trained animal acts as a restraint and helps calm the new animal. The yoke acts as a more severe restraint than a halter or lead rope, and the green animal quickly learns to follow the more experienced ox. The trained animal must be large enough to hold the unruly animal when he tries to escape. Some teamsters feel this method is the safest and most effective technique for training animals that are large and unruly.

Yoking the animals together and letting them loose to fight the yoke and learn to walk and move together is an option I do not recommend. This method was common in some areas of the United States and is still common in some parts of the world. This method of training works best after the oxen have fought each other and become weary in the yoke, and the teamster steps in. In the process of roaming unsupervised, the animals may learn to move about and do their own thing in the yoke, which most teamsters prefer to discourage. This method also requires a rugged yoke. Even so, the yoke may break, or one or both oxen may choke or be severely injured.

CHAPTER ROUND-UP

The early sessions of training older cattle can be frustrating, discouraging, and tiring. Spend plenty of time working with the animals individually in a small enclosed area to accustom them to being handled, so that when they are in the yoke they will be easier to control. Most teamsters use halters, ropes, and sometimes nose rings to maintain control during early training. The important thing is to maintain control at all times. If just once the animals learn to run away, the habit will be hard to break. ★

Raphael Santana
Santa Clara Province, Cuba

In 2004, I was part of a research delegation exploring Cuba's agriculture. My specific interest was in oxen, as I had heard the island nation had 400,000 cattle at work.

Most Cuban farmers had abandoned tractors when the Soviet Union collapsed and they lost access to spare parts and subsidized fuel. Fidel Castro then mandated that no bull calves on the island could be killed for a few years, and that farmers should return to using oxen, ensuring that people could grow food and other crops. This program worked to double the number of oxen on farms from 1990 to 1995.

We traveled around Havana looking at urban gardens, meeting with Ministry of Agriculture officials, and seeing historic sites along the way. A few days later our group finally arrived at a large farm cooperatively run by about 40 families in Santa Clara Province. Getting off our small tour bus, I saw my first oxen on our way into the community building of the agricultural production cooperative (CPA).

A group of farmers had gathered to meet with the Americans. There were formal introductions, and then the President of the CPA, Raphael Santana gave an overview through an interpreter of how the farm worked. Then the questions began. The Americans asked questions and the Cubans answered. Most of the farmers seemed less than enthusiastic about meeting with us. At

one point I had had enough, as I could see that the farmers, including Raphael, all wanted to be somewhere else.

I raised my hand and asked if anyone would like to see pictures of my oxen. There was an enthusiastic yes. The whole meeting quickly changed, from one of answering questions to real cultural exchange. Everyone got up and gathered around my oxen photos and the articles I gave Raphael. He proclaimed to all that this new information about oxen would be hung in the community center for all to see.

We then immediately left the building to go out and see some oxen. I could not have been more delighted.

As we walked down the country road, numerous teams of oxen and pairs of young bulls were passing us in head yokes. There were ox sleds and carts in the nearby driveway. I shared what I did with my oxen, and the Cuban farmers did the same. Even the interpreter said that it was the most exciting part of our week.

I learned that ox competitions are common, with one held right in the village during the month of July as an attempt to raise ox-training standards. Local schools included draft animals in their curriculum.

The standard in Cuba seemed to be to train young bulls at about age two for oxen. This surprised me the most. The calves that are left to run with the beef cows, I was told, are the wildest, compared to the calves of dairy cows, many of which were Holstein and Brown Swiss.

One pair of young bulls I saw in a yoke had already sired numerous calves, according to Raphael. The pair was dangerous, and I was told to not get too close. The trick to training this type of cattle in Cuba was to use the rings in the nose, chains attached to the rings, numerous trainers with large sticks, and then work the animals hard.

Sometimes the bulls were never castrated, and from the evidence I saw of one 10-year-old bull in a yoke, his battle had been long and hard. He had scars all over, his horns had been cut half off, and a nose ring had been ripped out and repositioned numerous times. Training mature cattle is not something for novices.

8
YOKE STYLES

okes have been used for centuries to harness the power of oxen. They are most often made of wood and are simple in design compared to the more complicated harnessing systems used for horses. Geography, cultural preferences, creativity, the availability of wood, and the task at hand all influence yoke design. Many yoking systems are crude, often reflecting the poverty and resources of the farm.

The first ox yokes were most likely head yokes used in southeastern Europe and western Asia. These crude early versions were not carved to fit the individual ox. Eventually they were customized for the animal, which maximized his comfort and his willingness to work. Yokes were first designed to be worn behind the horns, but a number of styles have been developed that fit in front of the horns, as well.

Great debate continues over the most appropriate yoking or harnessing system for cattle. Many factors must be considered. While most systems work, some work better than others. The most common system is a wooden yoke designed for two animals. Designs vary with local customs and regions of the world. An ideal system is one that minimizes breakdowns of both animals and equipment.

Importance of Fit

Whatever system of yokes or harnesses you choose, remember that animal comfort is the most important feature. Any design that does not fit the animal, or is crude and causes sores, is inappropriate. Far too many ox teamsters ignore the comfort of their animals.

A person wearing poorly fitting shoes quickly gets sore feet and is unable to walk. Even if the shoes fit, but are inappropriate for the task, problems develop. When the fit or adaptation to the system (whether shoes or yokes) is inappropriate, performance decreases and sores develop.

Not only must the chosen system fit initially, but the fit must also be monitored on a regular basis. Cattle grow quickly. They gain and lose weight. These changes affect the fit of the yoke or harness.

Oxen are most often worked in pairs, and the majority of this chapter will focus on the yokes

The Single Ox

Oxen are most often worked in pairs, and the majority of this chapter will focus on yokes that are designed for two animals. The single ox can be worked in a yoke, but this is less commonly used than the yoke designed for a team. The single ox yoke may be designed as a head yoke that fits behind the horns, a forehead yoke, or a neck yoke. Harnesses are also used on single animals.

The primary criterion in selecting a system to harness the power of a single ox varies little from the yoke or harness used for two animals:
- Make sure the yoke fits comfortably.
- Constantly monitor the fit.
- Use a system that is appropriate for the animal and the tasks for which it is intended.

designed for two animals. The single ox can be worked in a yoke, but this is less commonly used than the yoke designed for a team. The single ox yoke may be designed as a head yoke that fits behind the horns, a forehead yoke, or a neck yoke. Harnesses are also used on single animals.

The primary criterion in selecting a system to harness the power of a single ox varies little from the yoke or harness used for two animals: Make sure the yoke fits comfortably and constantly monitor the fit. Use a system that is appropriate for the animal and the tasks for which it is intended.

Harness

Most oxen around the world are worked in a yoke; few are worked in a harness. Sometimes the harness is designed specifically for cattle anatomy, and sometimes it is adapted from horse harness.

HORSE HARNESS

Horse harness is inappropriate for oxen. The carriage, gait, and movement of a horse's shoulders are different from those of cattle. Whether it is a breast band or collar style, the horse harness interferes with the ox's shoulders as the animal is working.

An ox holds his head much lower than a horse does, and his shoulder angle is much sharper and more prominent. The breast strap squeezes the shoulder of the ox, and the horse collar presses against the point of shoulder and must be lifted up with every step forward. Under a heavy load the prominent and flexible shoulders of the ox are severely restrained by a horse harness. Transferring a horse harness to an ox results in discomfort and poor performance.

Turning a horse collar upside down to raise the point of draft and use the ox's upper shoulders and neck may improve the comfort and success of the system. Yet with every step forward the collar moves

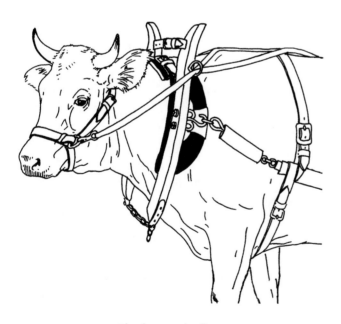

The three-pad collar.

right and left with the prominent points of the shoulders. As a result, an ox that is regularly required to plow or pull heavy loads wearing an upside-down horse collar will become sore.

While some ox teamsters claim to have successfully used the horse harness, the type of work the animal does and the frequency of that work greatly influence the success. Light work on a cart or wagon with a high hitch point may not create problems. Occasional heavy pulling may also appear successful. When oxen are asked to plow, log, or prepare for pulling competitions, however, the horse harness theory breaks down.

THREE-PAD COLLAR

The three-pad collar was developed in Germany at the turn of the century, with the goal of improving animal performance. If a harness is going to be used, this system — designed with bovine anatomy in mind — is probably the most appropriate. The three-pad system has the advantage of addressing the prominent and flexible shoulders of the ox and using the top of the neck and shoulders as the point of draft. It also allows animals that have weak horns or no horns at all, including cows, to work.

Compared to the yoke, this system is more expensive and complicated. A harness on an individual animal requires at least the use of a singletree and two chains for each animal. In multiple hitches an evener and additional chains are required. This system does offer more flexibility in harnessing animals of different sizes and strengths because it offers many possible points of adjustment.

Due to the limited experience of American teamsters with this system, and the lack of interest among competitive New England ox teamsters in adopting it, the three-pad harness would have to pass much scrutiny and many cultural barriers before being universally adopted for oxen in the United States.

Head Yokes

Head yokes were the most popular system in Western Europe, particularly in highland areas, and farmers in France, Germany, Spain, and other nations developed many designs and variations.

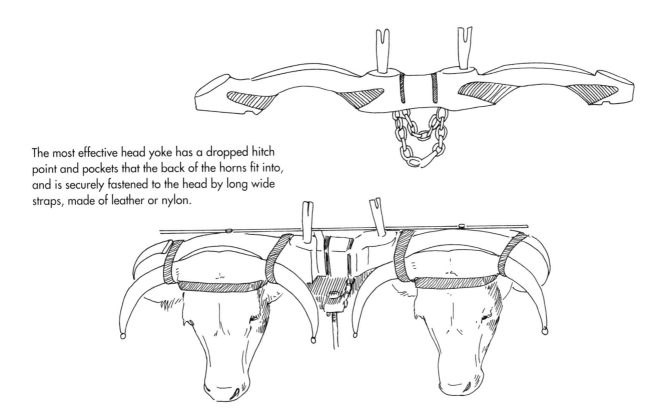

The most effective head yoke has a dropped hitch point and pockets that the back of the horns fit into, and is securely fastened to the head by long wide straps, made of leather or nylon.

As American and African nations were colonized, the early settlers brought their favorite yoke systems along with their cattle. Latin American farmers were influenced by Spanish and Portuguese settlers and used their versions of the head yoke. The French influenced the use of the head yoke in early Canada and in the French colonies of Africa. These cultural preferences continue today.

The head yoke fits behind the ox's poll and horns. To work correctly, the yoke must be carved to fit the individual animal. Every ox's horns are a little different from the horns of any other ox. For maximum performance, the head yoke pockets must be carved to fit the individual animal's horns. Fitting the yoke to the team is therefore a precise art compared to making a neck yoke. Once the pockets are carved the yoke cannot be used on any other team.

HEAD YOKE DESIGN

Similar to the neck yoke, the best head yokes are designed to hold the animals' heads slightly downward in a natural position. A good way to remember where their heads should be is to envision two bovines having a head-butting contest. Their heads are set slightly below the line of the back, with noses down. If the yoke pulls their heads too low (noses near the ground) they cannot generate maximum power. If their heads and noses are pulled into the air, the cattle are uncomfortable and unable to work effectively.

Pockets for Horns

The best head yokes have carved pockets to help hold each horn. As the yoke straps are wrapped around the horns and yoke, these pockets act like grips on the horns. They are carved with a chisel to be concave and slightly smaller than the horn itself. The edges of the pocket push against the outsides of the horns, while the straps around the horns pull them toward the pockets.

Do not have pockets that are too large or straps that are loose. The horns will roll around or slip easily in and out of the pockets and the yoke will not remain in a comfortable position.

Getting the animals to hold their heads correctly is a function of the dropped hitch point and of carving the pockets in the yoke so both animals hold their heads at the same angle. The lower the hitch point the more the animals' heads will be yanked down. If the hitch point is near the top of the yoke the animals' heads will be pulled up. If the pockets of the yoke are too large, allowing the horns to roll or move, the animals' heads will roll upward as they push into the yoke. Unlike the neck yoke, the head yoke should lock the animals' heads in the proper position.

Plenty of examples of head yokes can be found around the world, and many are sloppily tied to the horns with rope. Such a yoke might be a simple rounded tree trunk tied to the horns with rope or twine. A quick review of any picture of animals working in head yokes will indicate whether or not the yoke fits the animals. If they are not both holding their heads at the same ideal angle, the yoke or the fit of the yoke is incorrect. Like the poorly fitted neck yoke, a poorly designed head yoke may function, but it will soon lead to sore animals that are unwilling to give their best performance.

PROS AND CONS OF THE HEAD YOKE

The head yoke offers some advantages over the neck yoke. It provides more control and restraint. The animals adjust their heads both to lift and to push into a heavy load. Sores do not develop on their necks or shoulders from poorly fitted yokes. An animal wearing a head yoke will not get choked, as may an ox with a too-tight collar or neck yoke. On steep hills the head yoke allows oxen to slow or brake a load more easily.

The head yoke does have shortfalls. To get a good comfortable fit, the yoke must be carved to fit securely to the back of the animal's horns. Yokes that slip back and forth soon create discomfort and sores. The head yoke provides little flexibility in movement. On uneven terrain the animals must work with their heads tipped to one side or the other, increasing the chance of injury. The animals must have rugged horns. Oxen without horns cannot be worked in a head yoke, and oxen with small or weak horns run the risk of having the horns break

off. Yoking time is longer because each animal must be carefully strapped into its yoke.

In the northeastern United States, ox-pulling contests often include teams from nearby Nova Scotia, Canada, wearing beautifully designed head yokes. The Canadians stick by their design, saying they have more control and can generate more power because the animals have to pull together.

Year after year the debate over yoke designs is heard at competition pulls. Some years the cattle in head yokes win, and some years the cattle in neck yokes win. Both systems appear to work well in pulling contests on flat gravel surfaces with good animal comfort. The teamsters and the strength of the cattle have the greatest impact on who wins.

Neck Yokes

The most popular yoke in the United States has always been the yoke worn on the neck. This style is sometimes referred to as a "neck yoke," "bow yoke," or "shoulder yoke." The yoke rests on top of the neck and has bows to hold it in place. As the animal pulls, the bows are pulled into his shoulders.

The English and other farmers from northern Europe developed the neck yoke, although similar but more primitive yokes were used in Asia. The English experimented with many designs, including harnesses. English colonists influenced the adoption of the neck yoke in some of their former colonies, including the United States and Australia. Early American colonists used wood for both the yoke and the bows. Australian colonists also made yokes of wood, but often used steel bows.

WITHERS YOKE

Many African farmers did away with bows and instead used wooden staves or "skeis," with rope or a leather thong to hold the yoke in place. This style of yoke is called the "withers yoke" in Kenya, Tanzania, Uganda, and many other sub-Saharan countries, as well as in South Africa and India. Its specific design varies with culture, human ingenuity, and available materials. It is carried higher on the animal at the withers and uses different principles than the neck yoke, as described below.

When staves are used, contact with the mobile shoulders of the ox becomes a problem. Staves change the way the yoke works, because they push into the neck of the ox and interfere with his movement and comfort. However, when the hitch point is high on the yoke, as it is with the withers yoke, the staves work reasonably well on *Bos indicus* breeds. The yoke rests against the hump in front of the withers, and the staves do not interfere with the shoulders but instead turn forward away from them.

A well-fitted neck yoke sits in the middle of the animal's neck when at rest, easily rocks back and forth when the oxen move, and slides back into the shoulders when the animals lean into the bows.

Signs of Discomfort

Signs that the yoke is uncomfortable can include:

- The oxen throwing their heads up or down

- The oxen twisting their heads back and forth

- Reluctance to pull

- One animal not holding his head up

- One animal straining in the yoke, even under light work

If you ignore any of these signs your animals will not work effectively and may suffer long-term consequences, such as wounds, bruising, and scar tissue development.

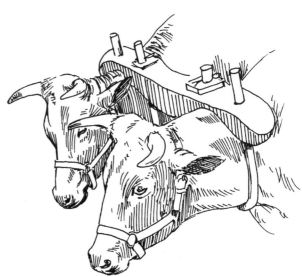

The withers yoke is designed to be used with humped cattle such as these East African Zebu. As the animals lean into the yoke, the yoke rides higher on the neck than the neck yoke does.

American Neck Yoke

The American preference for the neck yoke is admittedly culturally biased, but it offers a great deal of experience and possibilities for improving other designs. Unlike other yoke designs used in the developed world, the design of the neck yoke used in the United States has not been static for centuries. New England ox competitions challenge teamsters to maximize the comfort of their animals to get peak performance, so yokes are continually being improved and perfected.

Even within the United States neck yoke designs vary. The most common design is a single solid yoke beam with a dropped hitch point and carved neck seats to comfortably fit the animals. The yoke has holes drilled through it to accommodate the wooden

Neck Yoke vs. Head Yoke

Whatever variations are adopted, the neck yoke gives the animals substantially more flexibility and freedom than the head yoke offers. The neck yoke is therefore not as good a restraint or training aid as the head yoke, especially when the cattle are wild or unruly. The neck yoke's flexibility can create problems if the animals learn to turn the yoke, pull away from one another (haul out), or swing away from each other when asked to pull or back up.

On the other hand, the neck yoke may be easily transferred from one animal to another and is faster to put on and take off than the head yoke. For plowing, working on uneven terrain, and using cattle with no horns, the neck yoke works better than the head yoke.

	Head Yokes	Neck Yokes
Short heavy hauls	load easier to lift and start	better support of heavy tongue
Vehicle control	better on hills	yoke slides forward
Animal control	animals rigidly attached	more flexibility between animals
Animal comfort	less flexibility for animals	more comfort on uneven terrain
Animal flexibility	no	yes
Animal speed	slower	faster
Sore necks	no	yes, if yoke improperly fitted
Fighting in yoke	not possible	possible
Pulling away	not possible	possible
Horns	required	not required
Yoking time	slow	fast
Yoke construction	complicated	easy
Bow construction	none	appropriate materials required
Yoke fit	complicated	easy
Fits other teams	no	yes

bows, which are held in place with pins. Some ox teamsters use grab hooks or grab links on a chain to hitch the yoke to the load. Others have rings on the yoke to accommodate both a tongue and a chain.

In years past oxen were worked in multipurpose yokes designed for many jobs around the farm. Some teamsters had yokes designed for logging or plowing, usually the heaviest yokes, that brought the animals closer together. Other yokes were wider between the bows, designed for cultivating.

DESIGN CONSIDERATIONS

Some neck yoke designs are more appropriate and valuable than others. The value is not so much in the beauty of the yoke, but in how it rides on the necks of oxen hard at work. A beautiful yoke may be inappropriate when placed on the animals if it has been constructed without careful attention to

---

Slide Yokes

New England ox teamsters once used slide or slip yokes for animals that worked a lot on the road or in winter. The neck seats of this yoke slide right and left while keeping each ox an equal distance from the center. The system seems to provide more flexibility on slippery ground. It also helps prevent animals from hauling out, or pulling away from each other when they are in the yoke.

The slide yoke was the only neck yoke ever designed to have removable neck seats or neck pieces that could be changed depending on the size of the ox. The teamster therefore did not have to build a whole new yoke to accommodate a growing team, a new ox, or an animal that gained or lost a lot of weight.

Slide yokes were commonly used when animals were hitched to a cart or sled outfitted with a tongue. The yoke helped keep the ox on the inside of the turn from getting knocked over in a tight turn. Old photographs make it obvious that slide yokes were once quite common in New England. Many photographs are of "town teams" used for road maintenance and other work that a road crew might do today.

The slide yoke allows flexibility and comfort in movement, and offers the option of having one yoke with numerous different size neck pieces. The neck seats shown here are designed for 10-inch (on the yoke) and 11-inch bows.

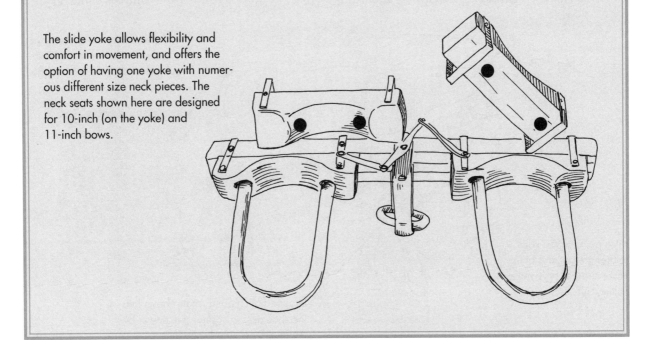

the animals' needs and the teamster's intentions. A number of factors influence yoke function.

Dropped Hitch Point

The neck yoke is designed to capture the power of the ox through his neck and shoulders. The design of the neck yoke has the hitch point at the bottom of the yoke. This hitch point acts as a lever, tipping the bows into the shoulders of oxen at work. The dropped hitch point pulls down on the animals' necks, forcing them to lift the yoke as they push into it. The lift aids in pulling heavy loads, such as might be encountered in logging, plowing, or pulling contests. If the hitch point were in the center of the yoke, the yoke would pull back onto the withers with little pressure against the lower neck and shoulders, decreasing the amount of power that could be captured from the team.

The dropped hitch point has other implications. When oxen are hitched to a wagon or cart tongue, the dropped hitch point does not function in the same manner as it would if the team were pulling something lower to the ground. The yoke will not slip into the shoulders, which can create difficulty in pulling a heavy wagon uphill because the animals can no longer use their power to lift the load to their benefit. Some yokes are designed with a hitch point that can be adjusted, perhaps by adjusting the height of the staple or by adding or removing wooden blocks from the bottom of the yoke where an adjustable staple is attached.

Staple

The position of the staple in relation to the center of the yoke is also important. If the staple hitch point is behind the center point of the yoke, it creates less leverage or pull, thus modifying the pressure on the shoulders from the bows. If the hitch point is ahead of the center point of the yoke, it will create more downward pull. Some modern yokes are designed with yoke rings or hooks that can be easily adjusted forward or back to change the "lift" the oxen have to apply to the yoke. The adjustment is based on the task, the implement being pulled, the animals' appearance, and their ability to function adequately in the yoke.

Neck Seat

I never realized how important the design of the neck seat and belly are until I put my first team into a yoke that was designed better than my own and saw an immediate improvement in performance due solely to a more comfortable fit.

Each side of the basic yoke should be uniform in weight, shape, and size from the center point of the piece of wood. Even if the animals differ in neck

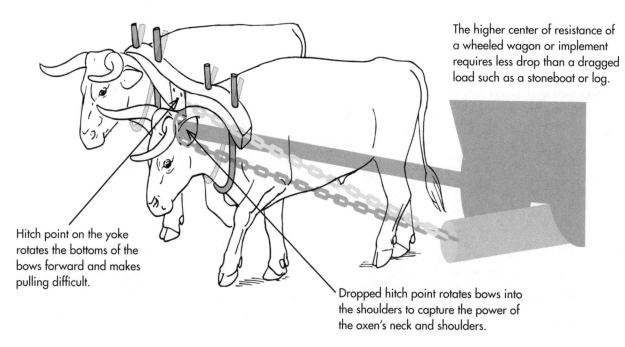

The higher center of resistance of a wheeled wagon or implement requires less drop than a dragged load such as a stoneboat or log.

Hitch point on the yoke rotates the bottoms of the bows forward and makes pulling difficult.

Dropped hitch point rotates bows into the shoulders to capture the power of the oxen's neck and shoulders.

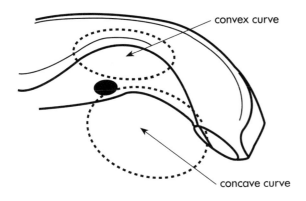

convex curve

concave curve

The neck seat is defined by two elliptical curves: a convex cross-sectional curve and a concave longitudinal curve.

size, the center of the neck seat on both sides should still be the same distance from the center of the yoke to ensure that the two animals do the same amount of work in the yoke. Occasionally the solid wood yoke beam will twist from one end to the other, especially if the yoke was constructed when the wood was green. This can create a situation where the two animals are working from neck seats that are no longer aligned.

The neck seat must have a slightly elliptical shape, horizontally both from right to left and front to back. Some teamsters prefer a flatter neck seat, giving the animals more surface area when pulling. The most critical part of the neck seat is where it presses against the animals' necks when they push into the yoke. The edge cannot be square or it will soon cause injury or sores. The cross-sectional ellipse must be comfortable for the animal.

The best shape for the neck seat is a question of much debate. Should it be flat, round, or elliptical? Many old New England yokes have a slightly elliptical shape to the neck seat, making it ride comfortably on the animal for a variety of tasks. If the curve from side to side and the curve across the front and back of the neck seat are brought together, they lead to an ellipse of sorts. If possible, try different yokes with various neck seats before choosing the design that works best for your team.

Smooth and Shapely

Because the neck yoke is designed to rest on the animals' necks, the neck seats must be carefully shaped and free of cracks, splinters, and rough spots. The smoother the surface the better.

The palm of your hand is a wonderful gauge of the neck seat's design and finish. Here's what you're looking for:

* A finished neck seat should be as smooth as glass.

* It should be uniform in shape from front to back and side to side.

* The two neck seats need to be similar and aligned on either side of the yoke.

* The neck seat must fit the animal comfortably.

Note: Do not pad this area of the yoke, because the padding may create sores as the yoke slides back and forth on the neck.

FITTING THE YOKE

The drawback of the neck yoke is that as the animals grow and develop, they will need progressively larger yokes. A teamster must always watch them

Fitting Bows

Cattle size	The bow fits if . . .
Under 1,000 pounds	A child's fingers easily slide between the neck and the bow.
1,000–1,200 pounds	The fingers of an average adult easily slide between the neck and the bow.
Over 1,200 pounds	The whole hand of an adult easily slips between the neck and the bow.

for signs of discomfort as the young animals outgrow their yoke. Oxen may be unwilling to work in a poorly fitted or improperly made yoke. A yoke will function properly only if it fits the team at every stage of their development and use. Every time you yoke a team, check the fit and carefully watch as they begin their workout.

As the ox wears the yoke, the bow is designed to slip comfortably between the animal's point of shoulder and the base of his neck. The position of the bow is critical to watch and understand. If the bow interferes with the movement of the shoulder because it is too large it will create sores on the shoulder. If the bow is so tight that it doesn't seat itself in this position, but stays up on the neck, the animal's skin will be severely pinched and become sore. A bow that fits too deeply will press against the front of the sternum or brisket. A bow that is pulled too high will choke the animal, making him cough, and cause sores to develop on the bottom of the neck. For many proponents of the head yoke, these are strong reasons to avoid using the neck yoke, but the simple solution is to make sure the yoke fits every time it is used.

Yoking a Team

Before yoking a team, tie both animals up to a fence or wall. Oxen that are accustomed to being yoked may stand without being tied, but tying them is a good safety measure. The animals should remain still until yoking is finished. If you are training unruly animals, yoking may require more than one person until the animals become accustomed to the procedure.

Yoking a Team, Step by Step

1. Over the neck of each animal place one bow so it will be readily available for the next few steps.

2. Place the yoke over the off steer's neck.

3. Holding the yoke with a leg (as pictured) or letting the nigh steer's side drop to the ground, turn the off steer's bow over and place it through the holes in the yoke.

4. Once the bow is in place, adjust it to the proper depth for the animal.

5. Add spacers to maintain the bow depth and place the bow pin through the hole in the bow.

6. Letting the nigh side of the yoke rest on the ground, call the nigh steer to step in.

7. The nigh steer should be trained to walk under the yoke as you pick it up.

8. Lower the yoke onto the nigh steer's neck.

9. Slide the bow into place on the nigh ox.

10. Position the bow, add spacers, and pin the bow in place.

CHAPTER ROUND-UP

Which system works best for harnessing ox power? This question has no single answer. Comparing harnessing systems is difficult. To make a valid test, the oxen must be accustomed and conditioned to both systems being tested.

Available resources, culture, and geography influence yoke design, as do the breed and size of the oxen. Be sensitive to but critical of new ideas. Remain will-ing to test them. The best test of a yoking or harnessing system is difficult or repetitive heavy work. If the animals break down, evaluate the comfort, design, and appropriateness of the system before giving up on it.

Always monitor and maximize the comfort of your animals. Whichever system you choose, physically condition your animals to the yoke or harness system and to the task at hand. ★

Nathan Hine
Worthington, Massachusetts

Nathan Hine grew up around oxen, showing in 4-H as a boy and later pulling cattle. Now he has his own family involved with oxen in Worthington, Massachusetts. His late father, Art Hine, was well-known throughout New England for his oxen, and his brothers Duane and Russell Hine continue to pull oxen at fairs all over New England.

Nathan grew up using traditional New England neck yokes, but about 12 years ago one of his Charolais oxen developed a sore neck, due to a bow not fitting right. The team had been pulling exceptionally well that season. Rather than give up on them, Nathan decided to try a head yoke.

He visited Mason Norton, an ox teamster in his 80s who had developed an interest in head yokes back in the 1950s. Mason had visited Nova Scotia, studied the yokes used there, and became the New England (if not the American) authority on head yokes for oxen. Many other folks who have used head yokes in New England, credit him for their success with head yokes. Nathan also acknowledged Hervey Ostiguy for helping a lot when he started, as he was the closest ox teamster also using head yokes.

Within days of putting the Charolais team in a head yoke, Nathan's oxen were working again. Some Canadian ox teamsters had told Nathan that it would take a few years to get the oxen back in pulling contests with head yoke. Yet within three months, the Charolais team were back winning pulling competitions. Nathan admitted he would make the switch again, if he had a steer in a similar situation.

However, when asked if he was convinced this was the system to use, he said, "No way!" It took him longer to yoke the team up in a head yoke than it did to harness up his horses. Also, "the yoke might as well be burned for firewood after the team is gone, because it will never fit another one. It is not as versatile as a neck yoke that could be used on many teams and last a lifetime."

He admitted that there were some advantages to the head yoke. "It gives you more control — what one steer does the other has to do," as the two animals are rigidly attached to one another.

Nathan has made several adaptations to the Canadian system. These include a totally adjustable yet stationary hitch ring on the yoke and a longer strap made of nylon, rather than the traditional leather strap used by Nova Scotian ox teamsters.

Ox yoke styles differ around the world. What ox teamsters know and use come from strong cultural traditions. Ox teamsters in Nova Scotia swear by head yokes, whereas most American teamsters prefer neck yokes. In competition, neither yoke style makes much difference in overall performance, but does make a difference in how the cattle are handled and used.

9

MAKING A
NECK YOKE AND BOWS

Although not difficult, designing and building an ox yoke does take a considerable amount of time. Some teamsters find it satisfying to design and create a yoke. Others prefer working oxen to manufacturing their own equipment. I have made many ox yokes, but because of time considerations and a lack of power tools I have also purchased yokes from professional yoke makers. Even so, I often take a sander, disc grinder, or saw to a newly purchased yoke to get exactly what I want. Many new ox teamsters may have no option other than to make their own yokes, or at least direct someone else in the process.

The Yoke

The yoke-making process includes measuring the animals to determine the yoke size, selecting the wood, selecting and tracing a design, carving the yoke, sanding, and finishing the yoke with some type of wood preservative. While you can easily make a yoke with simple hand tools, you can do many steps with power tools to reduce the time required to finish the yoke. The bows and hardware are more easily purchased than made.

YOKE SIZE

New England neck yokes are sized by their bows. Yoke makers use the distance measured horizontally between the bow shafts, when fitted in the yoke, as a beginning point in fitting a yoke to a team. This measurement is critical because it uses each animal's neck as a gauge of yoke size. Many experienced teamsters will be able to estimate the size yoke needed for a team just by looking at them. However, beginning teamsters may find ox calipers are handy for measuring the neck to determine the proper bow and therefore yoke size. The measurement is taken in the middle of the neck (as viewed from front to back), where the bow would be in a resting position when the animal is wearing a yoke.

The next critical sizing component is the distance between the two animals in the yoke. This is measured by the distance between the bow shafts to each side of the center of the yoke. After the bow size and the distance between the animals, the depth of the yoke belly and the width of the yoke neck pieces are the next most critical size components of the yoke. These components vary between yokes based on the yoke maker's preferences, the animals, and the use of the yoke.

SELECTING THE WOOD

Selecting wood for the yoke can be a challenge, since it is usually not possible to visit the retail lumberyard and find what is needed. Special-ordering wood may be possible, but the best woods for yokes

To determine the yoke size needed, measure the distance across the middle of the neck, where the bow rests when the animal is wearing a yoke.

are not readily available in retail outlets. Selecting a tree from the forest is the best option, but may not be feasible for many ox owners. Other possibilities include making a special order at a local lumber mill, talking to someone who owns a custom band saw mill, or asking a tree-service company to part with small logs for a nominal fee.

The best woods for an ox yoke are hardwoods that are difficult to split. Woods such as black or yellow birch (*Betula lenta* or *B. alleghaniensis*), elm (*Ulmus americana*), white oak (*Quercus alba*), hard maple (*Acer rubrum*), sassafras (*Sassafras albidum*), or cherry (*Prunus serotina*) are ideal. These hardwoods are not readily available in some areas, requiring investigation into alternative local varieties.

Numerous softwoods may be used for yokes, but their strength is limited. Cottonwood (*Populus tremuloids* and *P. grandidentata*) and similar softwoods make perfectly functional yokes, but they lack the strength required for some jobs.

The best log is free of knots and two to three times the diameter needed. Using a log large enough to quarter-saw or split out the yoke beam reduces the amount of center grain in the yoke. The center grain is the weakest part of the wood and the most prone to cracking. Although a yoke may be successfully made from a smaller log, the ideal yoke wood comes from a quarter-sawn or split log. Large-diameter seasoned hardwood is the most difficult wood to find.

AVOIDING CRACKS

Once the log is selected the wood must be prevented from splitting. While some amount of splitting may not affect the strength or integrity of the yoke, splits in the neck seat are hazardous to the animals and splits in other areas may weaken the yoke. Preventing splits presents a challenge, requiring advance planning, before the yoke making begins. The best wood for a yoke is a piece that has been seasoned for at least a year. Any cracks appearing by then are not likely to get worse.

Fast drying results in cracks. Putting green wood in a heated building, working the green wood in the hot sun, or even rapidly carving green wood on a dry day may result in cracking. Years ago, many farmers making ox yokes would bury the wood in a haymow and allow it to season before carving the yoke. Covering the wood while it dries still allows moisture to escape, but at a slower rate than if the wood were left in the open in a shed or building.

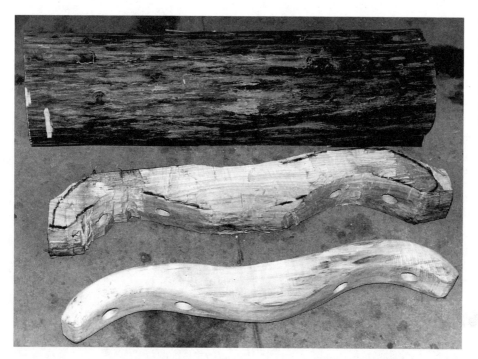

The yoke beam ideally should be split or quarter-sawn out of a large solid log. Minimizing the amount of heartwood in the yoke maximizes its strength and makes the yoke less likely to crack.

Drying wood may be painted or covered with linseed oil, especially on the ends of the log where evaporation will occur fastest.

. . . .

Years ago, many farmers making ox yokes would bury the wood in a haymow and allow it to season before carving the yoke.

. . . .

A yoke can be made from green wood that is quickly covered with a liberal coat of preservative such as linseed oil. This should be done after each step, in order to reduce the rate of moisture escape and prevent cracking. Some teamsters believe in making yokes from logs cut in the winter, when there is less sap in the tree, and quickly carving the yoke and covering it with polyurethane. The wood will dry slowly through the bow holes but may become dark because not all of the moisture in the wood escapes. Beware: these yoke beams may twist as they dry.

WOOD SIZE

The size of the animals to be yoked and the tasks they will do determine the size of the wood. The length of the beam should be about 6.5 to 7 times the bow width. For a 7-inch yoke, therefore, the total length of the yoke beam should be about 45.5 to 49 inches. If the animals are used almost exclusively on a cart, a longer yoke is better, because it allows the animals to make turns and maneuver the cart more comfortably. For plowing and logging, the heaviest kind of work, a shorter beam will bring the animals closer together and enable them to work more effectively. (*See* appendix II.)

TRACING THE PATTERN

After determining the size of your ox yoke, you must decide on a design. I recommend tracing the shape of a yoke that has been used successfully on other oxen.

Trace the yoke's pattern onto paper or cardboard and use the pattern to transfer or trace the design to a new piece of wood. Other options include drawing a design or buying full-scale drawings.

If you decide to draw a yoke pattern from scratch, I highly recommend the following design, which originated at Tillers International. Using the bow width as a guide, all the yoke's dimensions are a function of the bow width or the animal's neck, which is the gauge used to determine the size of the yoke to be built. While this system may not meet every need, it certainly prevents major problems in designing a yoke without a pattern or any sort of guideline.

As a general rule, the distance between the two animals should be 2.5 to 3 times the width of the animals' necks. If, for example, the animals are wearing a 7-inch yoke (7 inches equals the distance between the shafts of the bows), the distance between the animals in the yoke should be between 17.5 and 21 inches. Many teamsters have their own formulas. Bear in mind, however, that getting the animals too close together increases the chances that they will interfere with each other in the yoke. Having them too far apart causes them to seesaw back and forth at work, as they try to maintain their position in yoke.

Leave enough wood in the yoke beam to maintain its strength and integrity. Cutting a deep curve on the top of the yoke to reduce the weight and add to the shape may reduce the amount of grain that runs from one end of the yoke to the other, and maintaining the grain that runs through the wood is important to maintaining its strength.

Both sides of the yoke should be equal in size and shape. A good way to make sure they are the same is to design your pattern by tracing only half of a yoke. Use this pattern to trace lines on one side of the yoke beam and then flip it over to trace the other side of the yoke. Marking centerlines on the pattern is important to ensure that the two halves of the yoke line up with a centerline marked on the yoke beam. The pattern should have both a side view and a top view, including bow holes.

Sample Yoke Pattern

TOP VIEW

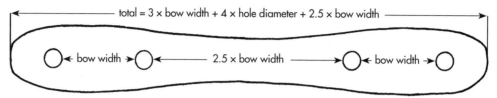

total = 3 × bow width + 4 × hole diameter + 2.5 × bow width

bow width · 2.5 × bow width · bow width

FRONT VIEW

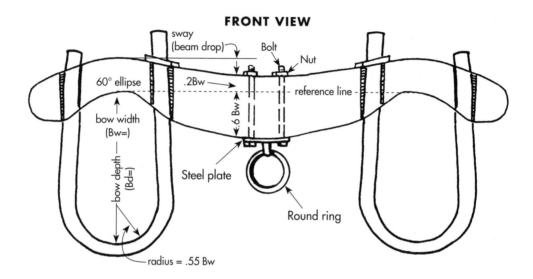

sway (beam drop)

Bolt

Nut

60° ellipse · .2Bw · reference line

bow width (Bw=)

.6 Bw

bow depth (Bd=)

Steel plate

Round ring

radius = .55 Bw

BOTTOM VIEW OF STAPLE

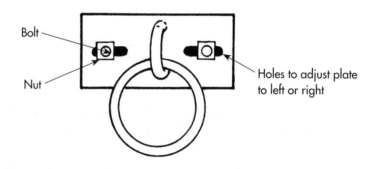

Bolt

Nut

Holes to adjust plate to left or right

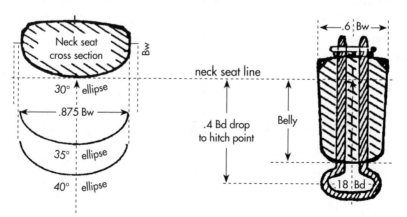

Neck seat cross section · Bw

neck seat line

.6 Bw

Cross sections of beams over the neck seats (left) and at the center (right)

30° ellipse

.875 Bw

.4 Bd drop to hitch point

Belly

35° ellipse

40° ellipse

.18 Bd

Yoke pattern designed by Tillers International, Scotts, Michigan

DRILLING BOW HOLES

Once the yoke beam has been squared with either hand tools or a saw, and the pattern has been traced onto the yoke, you're ready to drill holes. Whether you use a drill press or a hand drill with an auger, drilling is much easier when the yoke is still square. A square yoke beam allows the holes to be drilled more accurately than one that has been partially finished and wobbles on the workbench in multiple directions.

The diameter of the holes for the bows must match the size of the bows. The bow holes are drilled slightly larger than the bows themselves so the bows will slip easily in and out of the yoke. It is better to drill the holes correctly rather than shave the bows to fit into the holes, because shaving may weaken the bows. The bow holes must be as straight as possible and drilled in the center of the yoke. If the bows are off-center, one bow shaft may put pressure on the neck and shoulders of one ox more than the other, leading to problems. Off-center bow holes may also weaken the yoke.

SHAPING THE BEAM

New England has many ox yoke makers and each has a system for carving the log or beam into a functional yoke. Some use hand tools such as the adz, ax, and chisel to rough out the yoke beam. These tools work best in wood that is green and easy to work. The job can be long and tedious, especially if the tools are not sharp. Once roughed out, the yoke must be smoothed with a plane, drawknife, spoke shave, or rasp. Depending on the condition of the tools and the skills of the woodcarver, the amount of time spent on the yoke can be enormous.

Making their own yokes, many ox teamsters find the combination of a chainsaw and a rotary disc grinder with a large-grit sandpaper to be a feasible and inexpensive alternative to using hand tools. The disadvantage is that the chainsaw can quickly cut too much wood and can be dangerous as a wood-carving tool in the hands of those not familiar with its use.

In carving out the yoke, the fastest and most accurate method is to use a large band saw, thus reducing the chance of cutting too deep with an ax, adz,

or chainsaw. Using a band saw also eliminates a lot of the work involved with smoothing out the marks and deep cuts of tools that are less easily controlled. The only challenge with using a band saw is finding one large enough to do the job. Many woodshop owners with a less than adequate band saw have become disgruntled when the saw got stuck, burned out, or simply was unable to carve out a large yoke or one made of extremely hard wood.

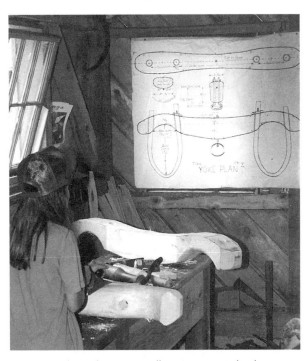

A young yoke maker uses a Tillers International yoke pattern as her guide for shaping a small yoke.

An ox yoke may be shaped with the simple hand tools found on nearly any small farm.

The most critical part of the yoke is the neck seat. During yoke construction, do not cut too much or too deep in the neck seat. It is critical to leave enough wood to get the right curves where the yoke will ride on the animal's neck. Carving the neck seat is often best done with hand tools or small sanders that are easily controlled. Although the visible yoke

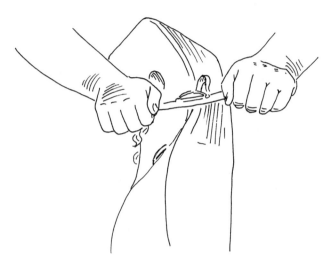

Make sure the neck seat is smooth and all the corners are rounded.

parts often receive more attention, extra time spent on the neck seat will pay off in comfortable oxen that are more willing to work.

Sand and smooth all flat surfaces and cover the yoke with polyurethane, linseed oil, or paint. This final coating protects the yoke from the weather, gives it a smooth attractive appearance and prevents any rapid moisture escape. Once the yoke beam is shaped and coated with a finish, its time for the bows and hardware. These yoke parts are constructed separately and require additional equipment and skills. You may prefer to leave these procedures to people with the skills and proper tools to do a good job.

ALTERNATIVE MATERIALS

Most ox yokes are made of wood, but numerous other options are available. Alternative materials include plastic, laminated wood, fiberglass, or combinations of all of the above.

Plastics

Given the large number of plastic materials available in the United States, ox yokes may be made of plastic, using injection molds. Depending on the plastic used, this option may offer additional strength without substantially increasing the yoke's weight. The kind of plastic used will have a bearing on the yoke's use. In cold weather plastics may become brittle and in hot weather may become flexible and may possibly bend.

Plastic yokes can be expensive because yokes are not produced in large numbers. Once the molds are made they may be used repeatedly, possibly at a lower per-unit cost. Plastic offers the advantage of not rotting or being destroyed by insects or wet weather. Although nontraditional, fully functional plastic yokes have been successfully used to harness the power of the ox.

Fiberglass

Fiberglass is sometimes used to cover yokes to protect them from the weather, insects, and rot. Fiberglass can create a smooth surface in the neck seat if the wood has cracked in that area. Filling cracks in the neck seat can also be done with one of the many wood filler epoxy products available. Fiberglass may be more brittle than wood or plastic and can be damaged if the yoke is abused or hit by a wagon tongue or other hard object.

Laminated Wood

A laminated yoke made of wood boards glued together may be stronger than a traditional solid wood beam, provided that strong waterproof glue is used and the boards are planed or sanded completely flat before gluing. The laminated yoke has some of the same disadvantages as a solid wood beam, such as damage from weather and insects.

On the other hand, the laminated yoke beam is not subject to cracking or weak spots created by splits and knots of a solid beam. The varying grain of each board reduces the chance of breaking or cracking. Laminating the yoke allows you to select the type of wood and the direction of the grain for the outside surfaces, creating opportunities for a beautiful yoke. Once the yoke beam is built, the steps in constructing the yoke are similar to those using a solid beam, including the need for a preservative finish.

The Bows

The basic technique for making wood bows is to steam and then bend the wood around a form to get the desired shape and size. It sounds easy. Anyone who has tried to bend wood knows better. Even the best bow makers wind up with a big pile of scrap wood at the end of the year. No two trees or pieces of wood are exactly alike, nor do they behave alike when forced under pressure and heat into an unnatural shape. There are many systems for making bows successfully, including bending bows without heat, using a fire instead of steam, and laminating thin strips of wood to a bow-shaped frame.

Other materials such as steel, pipe, or PVC have been used successfully on calf yokes or bows in larger yokes that will not be subject to the full force of a mature ox team with heavy pulls. These alternative materials will often bend when heavy pulling is involved. Alternative materials can also be troublesome during extremely cold or hot weather.

SELECTING BOW WOOD

Considering the job of the bow in an ox yoke, it is amazing that wood varieties exist that will withstand the force of a team of oxen pushing against the small diameter of the bow shaft day in and day out. In New England, only a few woods are considered strong enough for making bows. The favorite is the shagbark hickory *(Carya ovata)*, a dense and strong wood that is the top choice by most ox teamsters. Almost as good but not nearly as popular is the white oak *(Quercus alba)*. Even less commonly used, but functional, is the white ash *(Fraxinus americana L.)*. All three are strong, heavy woods with relatively straight grain and fibers, and resin that

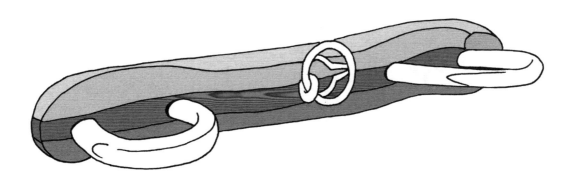

The laminated yoke may be easily made with boards and hand tools, resulting in an attractive yoke that is often stronger than a solid wood yoke.

Should You Make or Buy Bows?

Making bows is a lot of work. Great satisfaction can be derived from knowing that you can produce what you need for your team of oxen. At the same time, given a reasonable cost for a set of ox bows, you may be hard-pressed to justify setting up a steamer, building forms, finding the wood, and making and finishing the bows.

Many outstanding yoke makers buy their bows in New England, which offers numerous sources for purchased bows. These bows show much variation in style, quality, and the wood used. Before purchasing a set of bows, know your animals, what you will use the bow for, and what size and shape bows you need.

Most bow makers make bows part-time, and supplies and availability may vary. New bows may be hard to find in the fall (competition season), so plan ahead. Bow making is not difficult, especially when power tools are employed. Craftsmanship and competition are good for any industry, but it wouldn't take many bow makers to saturate the market because the numbers of oxen are limited and a well-made set of bows will last 20 or more years.

The best bows are split (not sawn) from the log and are made from the lighter sapwood (not the darker hardwood), as seen in this shagbark hickory.

loosens when heated. Woods in other areas may be equally appropriate.

Select a tree or a log that is free of knots, branches, and other defects. The most successful bow makers use a log 6 to 10 inches in diameter and at least 4 to 5 feet long. Individual pieces of bow stock are split from the log. The sapwood, not the heartwood, is used. Most hickory bows have the bark left on the outside, even when finished. This bark, which stays on the bow if the tree is cut during winter or early spring, serves two functions: it helps keep the outside of the bow from splintering when steamed and bent, and it shows that the bows were made from split rather than sawn stock. Split wood generally has a greater chance of the grain running all the way through the bow.

Ox bows are sometimes made of ash that was sawn from logs rather than being split. While such bows can serve the teamster, ash is not as strong as hickory. In the process of making the bows, a lot more skill is needed to select bow stock from wood that has been sawn rather than split from the log. The risk is far greater that the grain will run out of the stock, making the bow less strong than one that has been split and then whittled down to the appropriate diameter before bending.

PREPARING THE STOCK

Once you have selected and split the bow stock, bring it down to the appropriate diameter for the bow size you need. Use a drawknife or other appropriate wood-reducing tool. Be careful not to reduce the wood too much at this point. It is better to have the bow slightly larger than the diameter needed; you can always whittle off a little to make it fit into the yoke.

Before steaming and bending, the section of the stock to be shaped should be rounded and thinned, especially on the inside radius. Thinning the stock aids in preventing the wood fibers from splitting or pushing out as they are forced together.

Don't do too much finish work on the bow at this point. It might still splinter or break while bending. Complete the finish work after successfully steaming and bending the bow.

BENDING THE STOCK

You can employ a number of techniques to bend the bows. Steaming is an important part of the process. The steam heats the wood's resins, which are the natural glues that give the wood fibers their rigidity. If the wood has resins that respond to steam, the fibers will slip when the wood is forced to change shape, during bending. For most craftsmen, steaming the bow involves a closed, but not completely sealed, container: a pressure cooker that might explode is not a good choice.

A wood or propane stove can be used to heat a tank of water, which can then be connected via a hose or small pipe to a container where the bow stock is being heated and steamed. The bow stock does not need to remain in the container or box for long, provided plenty of steam is produced: 30 minutes to an hour is usually plenty of time. Some bow makers claim that with a good solid heat they can bend bows after heating for just 15 minutes. Most bow makers have steamers that hold several pieces of stock so they can bend one bow while others are being heated.

The hardwoods used for making bows are strong woods. Hours of steaming will not loosen the fibers enough to make them bend like a piece of rubber or assume any shape like pasta. Most require some kind of leverage to pull or push the ends of the bow stock around a preshaped form. The leverage might be supplied by a large reel, a come-along, or a hydraulic cylinder that forces the steamed bow stock to take the shape of the form. The form must be solid enough to withstand the combination of the strong wood and the force exerted upon it without resulting in a misshapen bow or a broken form.

Bow Shapes

Bows used in New England generally have a shape (at the curve) that is similar to the partial arc of a circle. Some teamsters consider this shape to be the most desirable for a bow. The most important aspect of making bows is knowing what your cattle require. The bow must fit neatly and comfortably into the ox's shoulders as he pushes against the bow. Some cattle are narrow in the shoulders, others are wide. Some have narrow necks like a cow, others are shaped more like a bull. Some animals are larger at the top of the neck, others at the bottom.

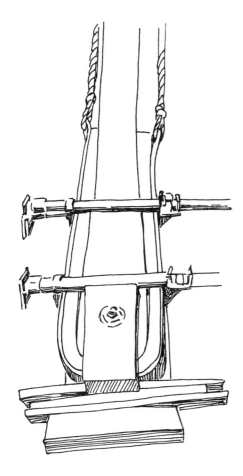

Homemade jig for bending a steamed bow, snugly placed in a metal frame to minimize twisting and splitting. Wedges driven between the steel frame and a block bolted to the bending jig keep the bow from coming to a point by slipping away from the form during bending. Cabinetmaker's clamps are used to pull the bow against the form.

Try Them On

Evaluate your animals to see which bow design is the most appropriate. Try different bow shapes on them during heavy work. If you study the animals' behavior during heavy work, you will clearly see the most comfortable bow shape for your cattle.

A comfortable bow is not so wide that it interferes with the movement of the points of the shoulder, nor so pointed as to cut off the wind or pinch the neck. If the bow is too tight the animal will twist his head from side to side as he tries to find a comfortable position in the bow. Once sore, the animal may become reluctant to work. Every ox requires a bow that fits him individually. If the two oxen in your pair have very different shapes, design your yoke accordingly.

When making bow forms, allow some flexibility in size and shape. Not every ox has exactly a 7- or 8-inch neck. The wood must be firmly placed against

A 9-inch and an 8-inch bow ready for storage: covered with linseed oil and tied with baler twine, with wooden spacers to help the maintain the shape.

the form and remain there throughout the process. If the wood shifts in the form, you might end up with a crooked bow that is not fit even to hang on the wall.

After a Successful Bend

Leave the bow in the form overnight, or longer if you don't have to reuse the form right away. If the form is needed to make another bow, take the first one off after an hour or two. In that case, tie the bow and brace the arc with a piece of wood so it does not pull in and form a point, instead of a maintaining a nice circular arc.

Once the bows have been shaped and are allowed to cool on the forms, they may be finished. Many bow makers use a special bench called a shaving horse that lets them sit and use both hands while holding the bow with a foot-powered clamp. The finished bow is smoothed with a drawknife, spoke shave, rasp, or sandpaper. Some bow makers have developed power tools to do this part of the job.

Caring for the Finished Bow

Just like any other piece of equipment or tack, a bow must be properly cared for. Here are some tips to make your bow last.

Sealing. Cover the bow with some kind of wood preservative or sealer. Most teamsters use linseed oil; some use polyurethane. A sealer helps prevent beetle larvae from attacking your bows and also prevents it from swelling if it gets wet. A bow that absorbs water will swell, which can result in its getting stuck in the yoke.

Storing. Store bows in a place that is dry and out of the sun. Most important, never leave them out of the yoke and stored untied. Left untied, bows that have been steam-bent will return to a straight piece of wood. The only wood bows that will not straighten are bows made from laminated wood strips.

Spacing. Use twine or wire to tie the ends of the bow shafts so they will remain approximately the same distance apart as they are in the yoke. If the wood is still green put a wooden spacer near the curved end to maintain the arc you have bent. A green bow will have a tendency to twist and turn as it dries.

Remove all splinters. Round all surfaces that will touch the animal, but be sure to leave enough wood so the bow fits snugly into the yoke and is strong enough to withstand the job for which it is intended.

The Hardware

Yoke hardware comes in many different styles and designs, all with the same function: to attach the animals and the yoke to whatever you want to pull. Basic hardware consists of a ring to which the load is attached, a staple that attaches the ring to the yoke, and bow pins that hold the bows in place on the animals' necks. The hardware should be strong enough so that it won't bend or break. It should be made with materials that are long-lasting and economical.

Beautiful hardware may be manufactured in a machine shop, but often at a great cost. Yoke hardware was traditionally made in a forge by a blacksmith. Such hardware usually lasted longer than the oxen and teamster, and even the yoke. Today many welding shops can put together yoke hardware that is both affordable and functional, combining modern technology with some of the techniques used by blacksmiths.

STAPLE AND RING DESIGNS

Yokes to be used for general farm and cart work may be designed with a simple staple and ring, a system that works well for two animals of the same size and strength. The size of the iron for making the staple and ring depends on the animals and the work they will be doing. Most hardware and yoke rings for mature animals are made of mild steel that is at least ½ inch in diameter. Rings and staples made of ⅝- to ¾-inch bar stock may be desirable for doing heavy work on a regular basis.

Some teamsters use a pair of rings consisting of one round ring and one grab ring, also called a calabash ring because its pear shape resembles that of a calabash gourd. When a pole is used to pull the load, the pole slides into the round ring; when a chain is used, it is slipped into the grab ring. The grab ring allows you to quickly and easily adjust the length of the chain, compared to passing the chain through the round ring and hooking it back into itself. When the chain is long, using a grab hook makes adjustments difficult. With a grab ring you do not need a hook on each chain. The grab ring must be large enough to allow the chain to slide easily through its round end, and must have a catch that will hold the chain fast, without slipping. The disadvantage to a grab ring is that it may not work well with all chain sizes.

The grab ring and staple may be made of square bar stock, but make the round ring of round stock because it is less abrasive on wagon tongues and poles.

Sometimes the staple consists of a steel plate fitted with rings and bolted to the bottom of the yoke with either two large bolts or four smaller bolts that run through the yoke beam. To accommodate oxen of different strengths, the plate may be made adjustable from right to left by cutting small slots instead of drilling holes, allowing the plate to be moved to one side or the other when the bolts are loosened.

Some yokes have no rings, especially if they are to be used exclusively for pulling with a chain. Instead of a ring, a grab hook is attached at the bottom of a plate on the yoke. During pulling competitions, this hook is easily moved to different positions to get the most from each animal.

The variations and designs for yoke hardware are endless and should not be limited by my suggestions. The basic idea is to have a strong and solid method of attaching a load at the bottom of the yoke.

Wraparound Style

Another common design (see next page) is to have a top and bottom plate bolted to the outside of the yoke beam, essentially wrapping around the beam. This design does not require drilling holes for a staple or bolts. It facilitates easy adjustments to the right and left, as well as accommodating blocks of wood placed under the plate on the underside of the yoke to lower the hitch point.

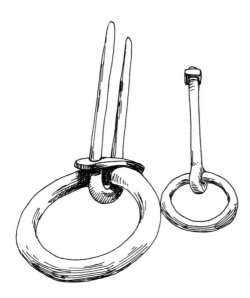

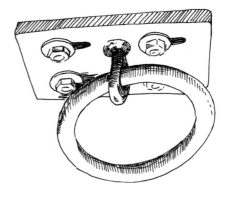

This steel-plate staple is bolted on using slots that allow the ring to be adjusted to the right or left to give one ox more leverage than the other.

Most older yokes have a stationary staple running through the center of the beam and fastened by a nut and washer or a steel pin; some hold a simple ring (top); others have both a round ring and a grab ring (bottom).

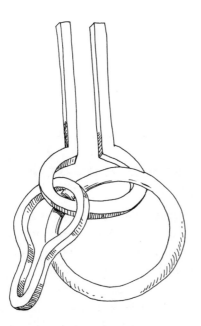

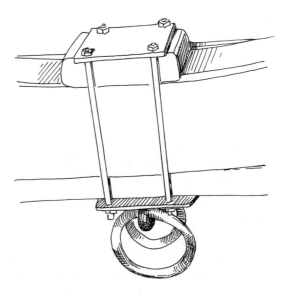

This staple is adjustable left to right, up and down, and front to back; lowering the hitch point and moving it forward pulls the animals' heads down and tucks the bows into their shoulders to improve leverage.

RING SIZE

The size of the yoke ring is determined by first evaluating how the ring will be used. A single general-purpose ring must be strong enough and large enough for any job the team might possibly tackle. It should be made of steel that will not bend when the team is plowing or pulling heavy loads. It should be large enough to accommodate any tongues or poles for pulling carts and other implements. The yoke ring should not be so big that the tongue or pole falls out easily or sways back and forth to any great degree.

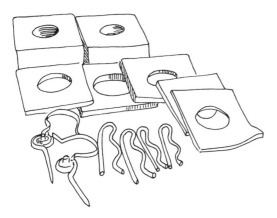

The most important feature of a bow pin is not its design or strength, but its ability to stay in place; wooden and leather spacers are used to adjust the bow's depth without the need for multiple holes.

For a small yoke designed for calves, a ring with a diameter of 4 inches is adequate. For larger animals the ring should be 4½ to 6 inches in diameter.

BOW PINS

Bow pins slip through holes drilled in the bows to hold the bow at the desired height, which may be adjusted by means of multiple holes or by using spacer blocks. The pins must be designed so they stay in place at all times. If a bow pin slips out and the bow falls out of the yoke, the situation could be dangerous for both the teamster and the oxen.

Bow pins are usually manufactured out of hardened steel. They come in many designs, the most common of which is similar in shape to steel pins used for modern farm equipment. However, the pins used for modern farm equipment may be appropriate only to hold the bows in smaller yokes. Pins made of hardened or spring steel are best because they can be designed to snap into place. Some bow pins are designed to slip through the holes and then wrap around the bow with small bands or fingers.

A blacksmith can easily make bow pins. Old hay rakes, tedder fingers, or even small springs may be heated and shaped into bow pins. Designs vary, but the function is always same.

The grab ring facilitates quick adjustments to the chain length and holds the chain adjustment better than a grab hook, which tends to disengage when the team is at rest.

Only one pin per bow is needed, as the bow is not under a lot of downward pull when the animals are working.

HARDWARE MANUFACTURE

Most yoke hardware is easily manufactured in a modern welding shop. With the use of torches and an arc welder, strong functional hardware may be manufactured quickly and easily. The welds must be strong and the steel large enough to do the job.

All yoke hardware was traditionally made in the forge, including welding the rings and making the hooks, staples, and pins. A good blacksmith may be hard to find, but the traditional method is still certainly a feasible option.

Some ox teamsters prefer to have their yoke hardware manufactured in a machine shop, which usually requires a shop specializing in custom work. The modern machine shop can manufacture yoke hardware out of a variety of modern steels and composites. Some of the materials available in a machine shop may not be available at a typical welding or blacksmith shop. Although the staple, plates, and rings can be made of the highest quality materials, the cost for manufacturing just a few pieces for an ox yoke can be prohibitive.

Yoke and Bow Care

An ox yoke is easy to care for and maintain. If it is used regularly in all kinds of weather, keep it covered with a good wood preservative. Frequently-applied linseed oil works well, as do polyurethane and paint.

The bows need regular care because the finish rubs off due to their continuous contact with the animals. The bows are the first things to swell in the rain because their end grain is directly exposed to the weather. As mentioned earlier, if the bows swell, they may be difficult to remove from the yoke.

When the yoke is not in use, always keep it out of the weather. Store it in a dry, clean environment out of the direct sun. Be especially careful of storing a yoke near windows or in vehicles that can heat up and cause cracking in even a seasoned yoke. Many cattle barns have too much moisture and too many cattle in them to be considered dry and clean.

• •

Never leave a yoke on the ground or on any surface where it might absorb moisture.

• •

A room or wall where the yokes may be hung or stored away from the weather and animals is ideal. Never leave a yoke on the ground or on any surface where it might absorb moisture. If a yoke is to be stored for a long time, coat it with plenty of wood preservative and cover it to protect it from dust and dirt. Properly cared for, an ox yoke will last for many generations.

CHAPTER ROUND-UP

The yoke is the most important piece of equipment used in working oxen. It must fit the animals comfortably for maximum performance. An improperly fitting yoke causes the animals to suffer as a human would suffer from wearing shoes that do not fit. Problems arising from a poorly fitted or poorly designed yoke may not be evident when the animals are used for light or occasional work, but when they are worked for hours or days, the weaknesses become readily apparent.

Using an ox yoke is easy. Monitoring its fit on the animals is the greater challenge. Finding or manufacturing and properly fitting a yoke are just as important as the training you give your team. Learn to read your animals in order to monitor the comfort of the yoke. Yokes that fit correctly will yield the best results with your team. ★

Tim Huppe
Farmington, New Hampshire

Tim Huppe is a rare individual who not only takes pride in his work, and has outstanding values and work ethics, but willingly shares all that he knows with others. He feels that his willingness to share with others has paid him back many times over, especially since starting BerryBrook Ox Supply.

Tim was a 4-Her in the early days of New Hampshire's working steer program in the 1970s. He and I share a common bond and history, but our years in 4-H never really overlapped. It seems like I have always known Tim, although I really got to know him when his daughters became interested in 4-H working steers. Having his four daughters ask, and even beg, him to join the 4-H program was a proud moment and a changing point in his own life.

With four daughters in 4-H, Tim fell into the yoke-making business. His daughters needed yokes, and the 4-H club needed someone to lead the club in this endeavor.

Tim's techniques and tools for yoke making are pretty simple: a tape measure, a pencil, a half round rasp, a hoof rasp, a chainsaw, an angle grinder, and a drill press. The keys are his attention to detail and a keen eye for a quality piece of wood. He developed this skill for finding the right wood during years in the lumber business.

His favorite woods for a yoke are yellow birch, black birch, red maple, and elm. "Yellow birch is wonderful to work with," he says, "and elm is desirable for the twist in the grain."

He tries to work with the piece of green wood as soon as possible after getting it home, and not to wait more than one week before making the yoke. Even so, he will always soak the ends of the yoke stock with one part boiled linseed oil and one part mineral spirits to minimize the cracking, as the piece begins to dry. Then once he really starts shaping the piece, he paints the entire yoke with the same mixture between each step, if he cannot finish it at once.

Before shaping the yoke, Tim draws out the rough shape and marks and drills all the holes while the piece is still square. He says, "I drill the staple from the bottom up, because you want the true center of the yoke at the point where the steers pull from. The bow holes I drill from the top down, as you end up cutting a lot of the material off the bottom of the yoke, in the neck seat." This way the bow holes are more perpendicular and centered from the top, where it is more critical.

Tim uses a chainsaw to shape the yoke, but says "a band saw actually eliminates a lot of the finish work with the grinder. I find the chainsaw easier when working alone, as the piece of stock (wood) can weigh 80 to 100 pounds and be hard to move on a band saw."

Tim's yokes reflect his style, and he spends a great deal of time and effort adding some finishing touches. The neck seat shape, rounded edges on the top and ends of the yoke beam, unique staple design, and his initials identify all the yokes he makes.

Tim says he has sold yokes all over the United States and his family has marketed working steers up and down the East Coast. He enjoys working with people old and new to the ox business and has had fun rekindling old friendships and making new ones with people interested in oxen. He hopes to soon expand both aspects of the business, and I expect both his steers and equipment will continue to make BerryBrook Farm well known to ox enthusiasts everywhere.

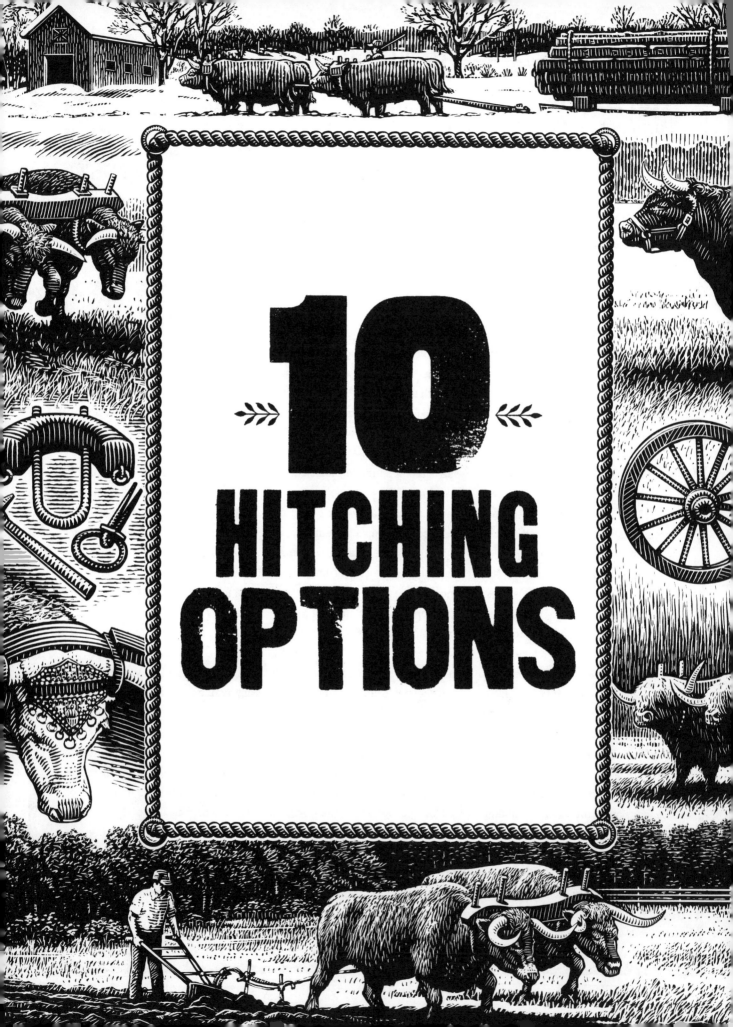

10
HITCHING OPTIONS

Most oxen are worked in teams of two. As herd animals, cattle have a calming effect on each other. Hitching a pair is easy, requiring only a yoke and single chain or pole. Two oxen may be readily walked in a straight line and controlled with voice commands. If one animal isn't paying attention the other usually is, making the animals responsive as a team.

The single ox is less common but may be found everywhere oxen are worked. The single ox offers an advantage for weeding between close rows and may be used for twitching logs and hauling a cart through narrow paths or roadways. Provided that the animal is large enough, a single ox can handle a plow or a cart designed for one animal.

Hitching multiple pairs together is no less common than hitching a single ox. The world over, two or three pairs have been hitched in tandem for plowing or logging. Teamsters once used six or eight pairs for hauling large wagons or huge logs in North and South America, Australia, and South Africa. Multiple hitches were essential on long journeys to ease the burden on each individual animal. Larger hitches consisting of numerous tandem teams working side by side and controlled by numerous drivers once moved buildings in both Australia and colonial America.

The Single Ox

Most breeds of cattle in the United States are large enough to be worked as a single ox around a small farm. The single ox will accomplish more work on a per-animal basis than will animals yoked in multiple hitches, and it is technically more efficient. The single ox cannot slouch or let another animal do all the work. Any work accomplished is the result of the individual animal's effort. For this reason, a single ox working alone will tire more easily than one worked in a team. Yet for the small farm with limited income or acreage, the single ox may be the ideal work animal.

If a teammate dies and a replacement ox is not readily available, the remaining animal may be trained to work as a single. An ox that is trained

to respond to all commands and knows how to pull makes an easy transition to working alone. All you need is a single yoke with two traces hitched to a singletree.

The challenge of working a single animal is maintaining control. A single ox tends to wander more in the yoke and is more likely to run away. Maintaining control is especially hard if a difficult-to-manage animal was formerly worked with a calmer teammate that acted as a stabilizing force.

For weeding, the single ox can maneuver in tighter spots, provided the rows are wide enough for the animal; steers or smaller oxen are needed for rows planted close together.

TRAINING THE SINGLE OX

The same rules apply for training a single ox as for training a pair, yet many teamsters who start out with a single animal take a more casual approach and demand less of the animal. These teamsters apparently view a single ox as more of a novelty than a real work animal. Provided the animal is trained with the same techniques used for training a team and is worked regularly, the single ox can accomplish most things a team can — just at a slower pace.

Many teamsters begin training a team by teaching each animal to respond individually to all commands before yoking the two together. Starting out by training each ox alone results in more valuable and flexible work animals. An animal trained singly may be yoked easily on either side, may be more easily hitched with an unfamiliar animal, and in case of the death of a teammate more readily makes the transition to working singly.

An animal that is trained singly and later yoked in a team should occasionally be yoked singly to keep him familiar with his initial training. Cattle are creatures of routine and habit. Mixing up the routine is important to get the animals to do any-thing required of them. Making an ox work singly, even after being assigned to a teammate, allows the flexibility of having animals that will work any way they may be needed.

Most oxen begin their initial halter training individually, but beyond that they are trained to function as a team. If a single trained ox is necessary for a particular task, identify one animal in the pair that appears more calm and trustworthy for use as a single animal.

Initially work with the individual animal in an enclosed area. Without his teammate, the single ox may try to walk away, but most well-trained oxen will quickly fall into the routine of working as a single animal.

As for training any animal, the earlier the training begins, the easier it is on both man and beast. A lead rope is important to initial training, but if the animal is to function at his best, don't let the lead rope become a crutch that you depend on. The animal must learn to be attentive, responsive, and willing to work with only your verbal, visual, or physical cues.

YOKING THE SINGLE OX

The single ox must be yoked using a system that is comfortable and designed with the animal in mind. If you use the bow yoke for animals in a team, then you should use the single bow yoke to work one

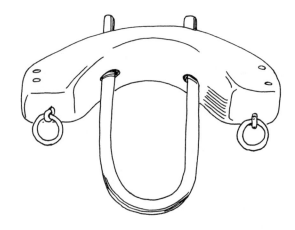

Single yoke with a dropped hitch point to prevent the yoke from being pulled up onto the withers, making the bows non-functional and possibly choking the animal.

animal alone. If the animal is accustomed to a head yoke, than a single head yoke should be used.

Because of the ready availability of horse harness, some teamsters think that it is the easiest way to harness the power of a single ox. Harness is designed for the horse's body, however, and is not suitable for the ox. Turning the horse collar upside down may work, by transferring the hitch point above the point of shoulder and having the narrowest portion of the collar at the shoulder. However, neither system works as well or fits an ox as a proper ox yoke does.

If you select the bow yoke, it must be appropriately sized so that the neck seat rides comfortably on top of the animal's neck. The yoke must be balanced to ride on the animal, so that it does not turn upside-down while the ox is working. The use of a britchen, which holds the yoke back against the shoulders, makes the yoke less likely to roll on the animal.

The hitch point of the single yoke is critical. The single yoke needs a dropped hitch point that pulls the bows into the shoulder pockets and forces the animal's head down slightly below the level of his back. This dropped hitch point is difficult to establish in a single yoke. With two hitch points on the yoke, depth adjustments are not easily made. The hitch points should place the traces out away from the animal's shoulders and body, so that they don't rub on the animal as he walks.

The single yoke requires traces and a singletree for fieldwork or two poles if hitched to a wagon, cart, or sled. You must place the two shafts far enough apart so that they do not rub constantly on the animal. The singletree must be adjusted so that it does not interfere with the animal walking — adjust the height or length so the animal will not catch his feet, pasterns, or hocks on the singletree.

WORKING THE SINGLE OX

Other than providing a single yoke and overcoming the challenge of initially controlling the animal, a single ox is suitable for a variety of jobs. The ox can be expected to respond as a team would to all verbal and visual cues and can be expected to work as hard in the yoke. The single animal will, however, have a tendency to move more from side to side on a heavy load.

For plowing, the single ox walks in the furrow. The animal cannot be expected to pull as large a plow as a team could, and requires more frequent rests. The single ox works well on a between-the-rows one-horse cultivator and adapts well to tools requiring the animal to walk in a circle, such as stationary grinders, pumps, and other animal-powered devices. The single ox can pull a two-wheeled cart, provided that it's well balanced or the weight is transferred to a back pad. The animal will tire quickly if too much weight is placed on his neck.

The single ox can also pull an appropriately sized wagon, sled, or stoneboat, and can work in places a team might not be able to go. The single ox can do

This yoke is too large for the ox; despite the dropped hitch point, the bow is too wide and too deep, and therefore does not to tuck into the shoulders as it should during cultivation.

A britchen (or breeching) helps keep the yoke back against the shoulders and less likely to roll.

almost anything a single horse can do, although he will do so at a slower pace and will not perform as well in hot weather. The single ox continues to function as he has for centuries, meeting the draft power needs of the small farm.

Multiple Hitches

As herd animals, oxen readily learn to follow each other. Working numerous animals in tandem is not difficult, as all the teams hitch into a single chain. The key to having a multiple hitch work is to get all the animals to start and pull the load together. When the animals are unfamiliar with working together, more than one driver may be needed initially. The disadvantage to working a multiple hitch is that individual animals learn to ease their workload by allowing the others to pick up the slack.

TRAINING MULTIPLE TEAMS

Pairs are more often hitched in tandem (one behind the other) than abreast (side by side). In a tandem

hitch the lead team must be responsive to all verbal and visual cues, and large enough to step out at a fast pace that keeps them ahead of the other teams. The more unruly or less well-trained teams are normally placed between the more responsive lead team and the wheel (back) team. The larger the hitch, the more important it becomes to control the less responsive teams in this fashion.

Large teams are difficult to manage at first because the animals do not know how to lean into the yoke all together, and they do not realize that they should follow the pair ahead of them. As the pairs learn to follow the lead team, the hitch falls easily into place. Teaching the teams to start the load and work all together can be another matter.

Stories abound of ox teamsters in the American forests who were called "bull whackers," "bull punchers," and other names that describe the cruel techniques used to get their oxen to start a load of heavy logs. The loggers used long sticks with nails on the ends to poke at the oxen as they ran by them. Stories tell of these bull punchers running down the

Teams that are first introduced together in a multiple hitch will work best with multiple drivers.

backs of oxen in caulked boots while screaming, to make the animals lean into their yokes all at once.

In any multiple hitch the efficiency per ox diminishes with every ox added to the team. Getting all the animals to lean into the yoke and work together is difficult. Even with four or six oxen, some animals purposely hang back and try not to work at all.

DRIVING A MULTIPLE HITCH

In a tandem hitch consisting of six or eight pairs, the lead pair and wheel pair respond to voice commands; the teamster stands near the middle of the team to direct the remaining pairs by voice, body movements, and whip.

Oxen in a multiple hitch will not respond to one driver in the same way as teams trained to work in pairs. In training a single pair, the teamster's body position becomes a cue to the animals as to what is desired of them. The same oxen put into a large hitch will try to position themselves with their teamster as they are trained to do — each will want to stand next to the trainer. The lead pair will try to

circle around and position themselves beside the teamster.

Training the animals to their position in a large hitch is like training oxen to do anything else — they learn what is expected of them through practice. Using multiple drivers is the easiest way to acclimate pairs that have not previously been hitched together. The animals must learn where they should be and how they should behave. First teach the animals to walk together in a line. Later, hitch them to something and have them learn to pull together.

HITCHING MULTIPLE TEAMS

Teams should never be hitched directly to each other's yokes. Animals can be injured if the team ahead yanks them forward too quickly. Hitching the animals yoke-to-yoke and attaching them to a sled, plow, or log near the ground creates a dangerous and uncomfortable situation. Only the oxen in front will be able to hold their heads up; all the animals behind them will have their heads pulled downward as the chains and yokes form as straight a line as

If the front team is hitched directly to either the chain or the rear team's yoke, the rear team's yoke will be pulled forward, possibly causing injury to the animals.

Hitching the shorter team behind the taller team is one way to better match the height of the chain to where the oxen prefer to hold their heads.

possible from the pair of oxen in front to the hitch point at the rear.

Hitching the tallest oxen in front and all the other oxen in a line back to shortest helps prevent this problem. The shortest oxen near the hitch point, however, will still have their heads pulled downward. The situation gets worse when the load is a plow, sled, or log with an extremely low hitch point. If the chains are hitched yoke to yoke, the team in front will walk along normally, while the last oxen in line will have their heads pulled down almost to the level of the hitch on the plow or sled.

By using short drop chains to hitch each pair to the continuous chain running from front team to the point of hitch, all the animals can pull while holding their heads in a comfortable and natural position. **I cannot emphasize enough the importance of using drop chains when hitching multiple teams,** whether you are putting together a long tandem hitch or just two pairs.

Since the chain running from the lead team to the point of hitch tends to form a straight line, each team behind the lead team should be hitched with a slightly longer drop chain so their heads will not be pulled closer the ground the farther back in the hitch they are positioned.

USING A VERTICAL EVENER

Tillers International developed the vertical evener for working multiple teams. It is especially helpful when hitching together only two or three pairs. The vertical evener works much like the evener used behind horse teams, except that it is not as large and it hangs vertically from the yoke.

Usually made of steel, the vertical evener serves two functions. First, like the horse evener, it is used to balance the load among pairs of animals of differing strengths. Second, it drops the hitch point of the chain running from the lead pair to load. The leaders do not use a drop chain or a vertical evener, but each pair behind the leaders require a greater drop to the hitch chain. The vertical evener is especially handy for pulling a plow or skidding logs, where the hitch point is extremely low.

The vertical evener allows the hitch chain to run in a straight line, and also adjusts for variable animal strength.

HITCHING OXEN ABREAST

Hitching oxen abreast is less common than hitching them tandem, due to the greater difficulty of hitching and controlling the animals. Hitching animals abreast requires an evener, numerous chains (which the animals are likely to get tangled in) and more complicated systems of control. One driver can see a large number of oxen in a straight line, but in a wider hitch the teamster loses some ability to direct the animals by visual or physical cues. For this reason, some ox teamsters hitching animals abreast use lines to direct the animals.

Teamsters have devised yokes to work three oxen abreast, more for training or show than for work. In order for a three-ox yoke to function, the animals must all be the same height. Hitching requires two chains and a singletree, and the hitch requires more room for turns.

CHAPTER ROUND-UP

The single ox is useful on a small farm for a variety of tasks such as row cultivating and pulling a light plow. Even if you have a team, working each singly adds to the animals' understanding of commands and makes them more responsive and flexible. Multiple teams are still used in numerous places around the world to accomplish hard tasks such as plowing heavy soils and moving large logs. ★

Lobulu Sakita
Lendikenya, Tanzania

I first met Lobulu Sakita in 1998, when I began my Ph.D. research in Tanzania. A Maasai man, Lobulu grew up herding goats, sheep, and cattle. He speaks three languages and had worked with other American researchers. We worked together daily, wandering all over northern Tanzania in search of Maasai ox farmers.

I discovered that Lobulu was not only good at conducting field research, but also an experienced farmer who has used oxen in the field since his childhood in Lendikenya, Tanzania.

Lobulu started using oxen as a boy. He recalled, "At the age of 10 to 12 years, I worked with oxen while plowing or pulling a sled or firewood. I started to drive the oxen when I was 10 years old, followed by holding the plow during plowing season and planting season."

I asked Lobulu to comment on hitching oxen and using them for farmwork. He replied, "When my family plows with oxen, typically we use four oxen, but also we use six oxen if there is a need. For instance, if the oxen are still young and the place we are plowing is hard, we use six oxen to increase more effort to the team." During the planting season he most often uses four oxen, because at that time the land has already been plowed. Everyone who has had oxen usually has a favorite, and Lobulu was no exception. Fondly recalling his favorite, he said, "Yes, I had an ox that was espe-

cially good at working in the yoke. What made him stand out as one to remember were his good habits. He would always obey and follow my instructions. Secondly, when he saw me arranging the yoke to catch the oxen for the work, he used to come alone to that place where the yoke had been arranged. It is very rare in our place to see an ox like that. Koroi [the ox] was my best ox, and I still remember him."

On the other hand, every ox teamster faces an ox he just does not seem to get along with. Lobulu offered the following comment: "Yes, we had an ox that would not work well or sometimes not at all in the yoke. In 2002, I had one ox I tried to train, but he completely refused to work. He used to lie down, refuse to walk, and not obey my instructions, and finally I decided to leave him behind and train another one."

Lobulu and I worked together for about six months in 1998 and 1999 in Tanzania. In 2002 we traveled to Kenya and then to Uganda, to attend a conference on draft animal use in agriculture. I asked him if he has changed the way he does things,

after seeing how oxen were used in other areas.

He replied, "Actually, I have made some changes with oxen since we began our work in 1998 to 1999. During that time, I used oxen for plowing and planting. More recently I am using oxen in various activities, such as transfer of the crops from the field to the homestead during the harvest season, and transfer of water from the water point [wells] to my home area. I also use the oxen to transfer manure from the boma [his farm] to the field. These things I did not do before we started doing our work."

Unlike many teamsters featured in this book, Lobulu has to train oxen in order to provide food for his family. His family truly depends on oxen, as tractors are not a viable option. Working with Lobulu was one of the most exciting things I have ever done, as we saw and met many ox teamsters who truly depended on oxen for their survival. We also had many adventures together, which are stories we both share with our family and friends. I found a lifelong friend and someone for whom I have developed the highest respect.

11

OXEN
IN
AGRICULTURE

espite a rise in draft-horse use among small farmers in the United States during the latter part of the twentieth century, the use of oxen in agriculture in the United States remains virtually unknown. The heyday of oxen in this country was before the Civil War. Prior to that, in the seventeenth and eighteenth centuries, the ox was the largest draft animal that could be found and put to work.

As the large European draft horses were imported, any advantage in the ox's size and strength was lost. At the same time, the ability of oxen to consume coarse native grasses, coupled with their low price, became less important to farms that were rapidly expanding in size and profitability.

The ox remained in use longest on poor farms where horses were not affordable, and in New England, where a preference for cattle had a strong cultural heritage and geographical advantages. Farmers eking out their living on small farms with rocky soils preferred oxen for plowing. Being slow, oxen were less likely to break equipment on large stones in the furrow.

The diverse nature of small farm operations provided the ox with year-round activities including plowing, hauling maple sap, haying, harvesting, logging, and rolling snowy roads. New England farmers appreciated the ox for his willingness to work in deep snow and mud and for his ability to survive on the poor-quality forage and other crop by-products

that were often his only feed. The oxen-powered farm in the United States, however, is now largely a thing of the past.

Farming with Oxen

Why would an ox be employed in any farming operation today? Unbeknownst to many Americans who use draft horses or mules, working bovines elsewhere in the world outnumber equines four to one. The vast majority of the world's small farmers find the ox to be the only appropriate animal for their agricultural operations.

. . . .

Globally, working cattle outnumber horses four to one.

. . . .

Horses cannot survive in many of these regions because of debilitating diseases and parasites. As

Plowing with oxen is as common today in Tanzania and other parts of the world as it was in North America 100 years ago.

they were for early American farmers, horses are out of reach for many small farmers in the developing world because of the cost of purchasing, feeding, and harnessing them. It is completely inaccurate to think that the use of oxen in agriculture is insignificant. They remain the most common draft animals in the world.

If oxen are going to be used for farming in more developed countries, however, there must be a good reason. Tractors, horses, and mules are faster. Using oxen requires a special commitment to adapt farm implements to the animals' unique way of working. Some farms, though, might be able to capitalize on ox power as a marketing tool. Others may use oxen to create a more sustainable small farm system. Museums have a reason to use oxen, as do institutions where international development workers and Peace Corps volunteers are trained to use them.

With training and physical conditioning, the ox can do any job on the farm. Animals that are accustomed to working long hours are easily put to work in any agricultural situation. Most oxen in the United States today are idle much of the time. Too many teamsters expect their oxen to behave like work animals, but only when it is convenient or fun to work them. Work animals need work. Regular exercise throughout the year is the best preparation for farm work.

The time commitment for using oxen is greater than for tractor use. Some hobby farmers use oxen only for certain tasks that require little prior conditioning. Raking or tedding hay, hauling manure, and other jobs using wheeled implements are easy on the animals.

Plowing, on the other hand, is one of the most physically challenging farm jobs for an ox team. Cultivating row crops requires the highest degree of animal control. The animals' limits depend on keeping the oxen in shape and in comfortable well-fitted yokes. Selecting implements that are appropriate for the task and the animals ensures a positive outcome.

Little has been published on the use of oxen for agriculture in the United States. Opinions about how much they are capable of doing, how they are being used, and how they may be adapted to modern operations are based largely on speculation and a few examples. So few examples of the use of oxen in this country exist that the easiest way to learn about oxen in agriculture is to study their uses in the past or adapt what has been written about their use in foreign lands and developing countries. The next best thing is to study draft horses in agriculture. As long as the limitations and yoking of the ox are taken into consideration, many of the same principles apply to both oxen and horses.

Training steers for farm work was one of the most enjoyable chores for a farm boy like this fellow from South Weare, New Hampshire, in about 1894. The young team learned its way around the farm long before they did any heavy work.

DRAFT-HORSE TECHNOLOGY

Few farmers understand oxen, and even fewer pieces of farm equipment have been designed specifically with bovines in mind. Most farmers who used oxen early in this century adapted equipment manufactured for horses. The ox is slower than the horse, and his simple yoking equipment may have to be adapted from horse-drawn implements. Any implement with substantial tongue weight will soon fatigue the animals unless a back strap or additional wheels are added to support the implement.

Hitching oxen to equipment is usually faster and easier than hitching other draft animals. Eveners or singletrees can usually be removed from implements and replaced with a simple clevis or single hitch point. The simple ox yoke requires few adjustments beyond a securely hitched chain and implement tongue. Simply running the tongue into the yoke ring and attaching the animals by means of the chain is the most common system of hitching.

The easiest way to adapt implements designed for horses or small tractors is to use a basic forecart. A forecart is a wood or steel pole attached to an axle that runs on a pair of tires. The forecart must be equipped with a sturdy hitch and be able to carry the weight of the implement it is hitched to. Hitching an implement to the forecart is as easy as dropping a hitch pin through a hole. The forecart is essential for serious farming with oxen. The combination of animal power and a small internal combustion engine mounted on the forecart offers great potential for farming with oxen by meshing animal power and modern equipment. A three-point hitch and a power take-off on a forecart may be used for most small pieces of modern agricultural equipment.

Oxen trained to follow directions from the teamster walking beside the animals can also be driven from behind with lines or voice commands. Adapting draft-horse equipment, forecarts, and implements to oxen makes their employment more efficient; however, motorized forecarts and modern implements represent a major investment, although plenty of old pieces of equipment are available for a reasonable cost. Save your money initially until you see whether ox power will be useful on your farm.

OX FARMING RESOURCES

To use an ox in the United States today requires more radical thinking and creativity than is required for farming with horses or mules. No manufacturers make recommendations for oxen and few resources are available for people needing help. In the past, New England farmers often employed local artisans to design carts, sleds, and other equipment with the ox in mind. That same creativity may be needed to use the ox today.

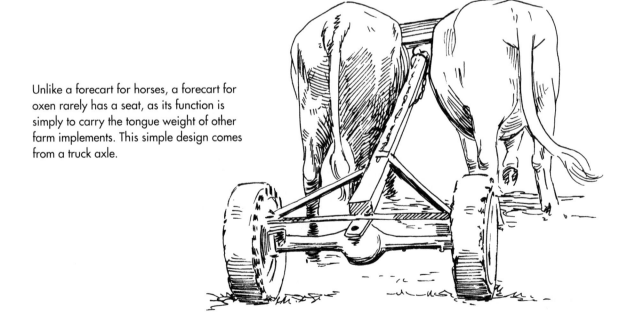

Unlike a forecart for horses, a forecart for oxen rarely has a seat, as its function is simply to carry the tongue weight of other farm implements. This simple design comes from a truck axle.

Ample written material is available on the use of horsepower in modern agricultural operations. *Rural Heritage* magazine and *Small Farmer's Journal* (see Resources) offer regular sources of information on using draft animals in modern agricultural operations. Both regularly feature articles on adjusting and maintaining equipment and incorporating innovations into the draft-animal-powered farm. While these magazines contain many articles about the draft horse, their information can be readily applied to oxen.

The following discussions evaluate how oxen were used in the past. Using your ingenuity, you can likely find better ways of farming with your oxen.

Clearing Land

One of the first jobs of oxen around the world has been clearing land in preparation for growing crops. Oxen can move brush, haul timber, pull stumps, and break tough sod.

In the early development of the United States, oxen helped clear land to make way for agriculture. The time and effort expended is often overlooked in this era of powerful machinery. Many modern homes are built in agricultural fields because it is too time-consuming and costly to clear forested lots. A teamster with the time and willingness can still tackle the job with oxen. Clearing land is slow and tedious. The ox is well suited to the pace and power required for such work.

LOGGING

Land clearing often begins with removing trees. The larger the trees, the greater the number of oxen required for the job. Loggers can cut the timber to a size that matches the ability of the animals to pull, and they can use sleds, carts, and wagons to ease the burden on the animals. Logging rivals plowing as one of the most difficult jobs on the farm.

Winter is the most appropriate time to log with oxen because obstacles like stumps and rocks are often buried in snow, and the frozen ground offers more solid footing. Cooler weather also provides a more comfortable working environment when oxen are moving large logs.

PULLING STUMPS

Once the logs and firewood are removed and the brush is burned or piled, removing stumps is the next requirement in clearing land. Pulling stumps is a time-consuming and difficult job. Attempt to pull only stumps that are a few feet tall. The roots will come out more easily if you can hitch the stump near its top, creating a lever action.

Early farmers in America cleared forestland and grazed livestock around the stumps until they had the time and equipment to remove the stumps. Many fields remained heavily pastured until the stumps rotted. Leaving stumps in place for a year or two allows the roots to partially rot, making them easier to break free.

Clearing land for agriculture is an important chore for this young Maasai ox drover in Monduli, Tanzania.

Pulling Stumps Properly

Never attach oxen directly to a large stump or you may make them reluctant to pull. A stump is firmly attached to the ground and requires a lot of power to break free. Using oxen to pull the stump by trying to jar it out of the ground is hard on the animals' necks and shoulders. Attempting to pull a stump without mechanical advantages will lead to broken equipment and a team that becomes soured by the task.

Over the years a number of devices have been designed to increase the pulling or lifting power of oxen. Pulleys, a block and tackle, levers, or lifts can help yank the stump out of the ground. Historically a team pulled stumps by turning a huge wooden screw by means of a long lever attached to the yoke. The stump was twisted, lifted, or torn out of the ground.

Various devices for pulling stumps

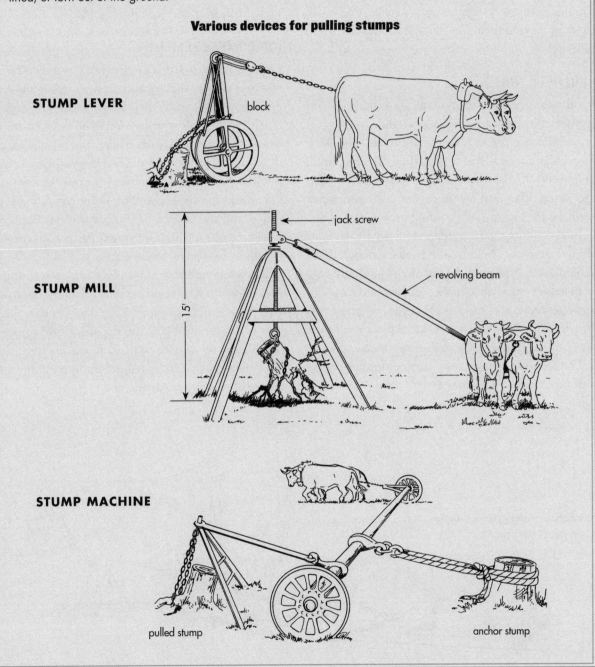

STUMP LEVER   block

STUMP MILL   jack screw   revolving beam   15'

STUMP MACHINE   pulled stump   anchor stump

Once cut, coniferous trees do not usually send up shoots, so they are not likely to survive heavy pasturing. They rot within 8 to 10 years. Cut these tree stumps close to the ground and they seldom create problems in a rough pasture.

Hardwood stumps tend to grow numerous shoots, often creating a stand of brush that is thicker than the trees that were cut down. Removing hardwood stumps rids the pasture of this possibility. Stump removal was always a huge burden that was eased with mechanical devices, most of which came much later in agricultural development and at great expense.

HAULING STONES

An abundance of stones in many areas makes farming difficult. Removing stones is tedious work that is made easier through the use of oxen. The New England stoneboat was designed for hauling small stones out of fields after primary or secondary tillage. Large, flat, and low to the ground, stoneboats are ideal for loading and hauling stones.

Larger stones or boulders had to be dug or blasted out before being loaded onto stoneboats and hauled away. On many farms the stones gathered from fields were used to make walls around the field. Since stones could not easily be moved great distances, stone walls were a perfect solution for disposing of them. Enduring for centuries, stone walls can still be found in many a rocky agricultural area that was once cleared with oxen.

Building Roads

Moving things around the farm or to the market required roads, for wagons or carts in warmer months and sleds in winter. Road-building generally begins by clearing a path over areas that are not full of daunting obstacles like boulders, swamps, lakes, ponds, and rivers. Many modern New England roads follow paths laid by early settlers and their ox teams. Such a road traveled the path of least resistance instead of the straightest path.

ROAD IMPROVEMENT

Once a rough road was cleared, it was easiest to travel in winter when swamps, ponds, lakes, and rivers were frozen, and when snow covered any irregularities or small stumps along the way. In many areas snowy roads were rolled with huge oxen-drawn snow rollers that packed down the snow and allowed sleds to pass for many months of the year. Oxen wearing shoes could draw enormous loads on a sled and travel great distances without the heat, insects, and mud that accompanied early wheeled transport in the warmer months.

Traveling on those early roads was slow, bumpy, and difficult. For much of the year many roads were more easily traveled by sled than by cart. Mud was a serious problem during the rainy seasons. Stoneboats or wagons with wide wheels were welcomed on roads as they helped smooth out the ruts created by carts and sleds.

A stoneboat is designed for hauling stones out of the field.

Later roads were improved by placing logs crosswise in wet spots or low areas, creating corduroy roads. They in turn were improved with planks and were called plank roads. These roads functioned well in winter and were usable in warm months, but eventually the rotting logs or planks made them impassable or in need of repair.

Stones were plentiful in most areas of New England and offered a permanent solution to traveling over wet and muddy areas. Oxen hauled large stones by stoneboat and smaller ones by cart, and the stones were dumped into low spots. Slip scrapers and road scrapers pulled by animals were also used to move soil or sand to the low spots and to ditch the sides of the roads.

Permanent roads were built and maintained through the use of oxen and later with other draft animals and heavy machinery. Many New England towns allowed taxpayers to pay their tax bills by maintaining and improving roads with their animals throughout the year.

Oxen can still be used to create a simple farm or woods road and to level a gravel road with a simple road scraper. They can pull road graders designed for horses. Ditching the sides of roads, hauling in new material, and filling in low spots with stones may be done using dump carts or stoneboats. The best time to work on a road is just after the rainy season. The road will be moist enough to work some

of the gravel loose without loosening the roadbed to the extent that it washes away in rainy weather.

SLIP SCRAPERS

Before the bulldozer and backhoe, the ox and the slip scraper were an alternative to shoveling. The slip scraper is designed to dig when the handles are lifted and to slide across the ground when the handles are left alone. To dump the load, the handles are given a quick lift, causing the front edge to dig into the ground and the scraper to flip over. Most slip scrapers were made of metal, although some were wooden with metal edges.

Provided the soil was not too rocky, the slip scraper worked well, allowing oxen to move soil for

To make travel easier for sleds, in northern areas snow was packed on roads with snow rollers like this one in New Hampshire around 1896.

short distances with little human effort. Soil that had to be moved a great distance would have been shoveled into a dump cart or wagon. When long narrow ditches were dug, a single animal was used because of the difficulty for two animals of walking in an uneven, narrow ditch.

Most slip scrapers were not large because repetitive digging and moving large amounts of soil would have quickly tired the animals. Instead, most were about 3 to 3½ feet wide and 3 feet long with sides about 1 foot high. This size allowed the scraper to haul about ⅓ cubic yard — an amount of soil, sand, or gravel a team of mature oxen can readily move all day long.

Primary Tillage

Primary tillage is a common job for oxen in many nations. Plowing is the most common form of primary tillage, with the following benefits:

★ Plowing tears up and buries organic trash, manure, or sod and helps eliminate weeds.

★ It aerates the soil and places organic matter into the root zone where it will rot and provide nutrients to growing plants.

★ It may also reduce soil compaction on the surface.

The baring of the soil by plowing is followed with secondary tillage, which provides seeds with good soil contact.

PLOWING

Plowing with oxen takes many forms. In some parts of the world the plow is the only implement on the farm and plowing is the oxen's only function. In some areas plowing is accomplished by scraping the surface with a simple pointed plow or ard to provide soil-seed contact.

The moldboard plow was a great invention and has been improved over its centuries of use in Europe and America. Yet despite the amazing ease with which a well-scoured and -adjusted moldboard plow can turn a furrow, plowing is still the most difficult task for a team of oxen. The physical challenge of plowing is made more difficult by the many variables affecting the plow's movement through the soil.

Conditioning Is Crucial

Before using your animals for plowing, physically condition them for the job. If they are not physically prepared they will have great difficulty following commands, staying in the furrow, and walking a straight line. Plowing is difficult even for animals that are worked year-round. Animals that are idle for much of the year will benefit from three to four weeks of prior lighter work like logging or pulling

A slip scraper slides across the ground when the handles are left alone (upper drawing). It digs or dumps its load (lower drawing) when the handles are lifted.

a heavy stoneboat loaded with concrete, stones, or other heavy objects.

Pulling a cart is much different from plowing and will not provide adequate conditioning. The early colonists always had their animals ready for work in the field by using them for logging in winter, hauling maple sap in spring and hauling manure to the fields prior to plowing.

Plows

Understanding the plow is as important as conditioning your animals. Learn to maintain it, and how to adjust it for different soil conditions. More types of plows probably exist than breeds of cattle to pull them. Talk to other ox teamsters and see what works for them. Better yet, volunteer at a farm museum with many working plows and try a number of designs. Choose one that fits your team, your soil conditions, and your budget.

Walking plows are the most common plows used with oxen and are relatively simple to adjust and maintain. A walking plow has few parts, but each part has an important function. Learn the parts of your plow and how to recognize when they are broken or worn.

The off ox must learn to walk in the furrow.

Riding plows are less commonly used with oxen. They are a feasible alternative provided they are maintained and properly adjusted. Many sulky plows were designed for horses and have to be retrofitted for oxen. This undertaking is not usually major. If you have a plow designed for draft horses, bear in mind the number of horses required to pull it. Do not short-change your oxen by trying, for example, to hitch two young steers to a plow designed for two or more mature draft horses.

Plow Adjustment

Walking plows are adjusted by changing the hitch point and the chain length. Ideally the plow design should enable both vertical and horizontal hitch point adjustments. Properly adjusting the plow is essential. Plowing should not be a struggle for either the animals or the teamster. The plow should ride easily into the soil, and any adjustments should be made with little effort.

The horizontal adjustments determine the width of the cut. Adjusting the hitch point toward the unplowed ground increases the width of the cut. Adjusting the hitch point toward plowed ground narrows the width of the cut.

The vertical adjustment and the chain length, which both control the chain angle, determine the depth of cut. The shorter the chain, the more lift on the plow and the less the depth of cut. A longer chain creates a lower angle and less lift on the plow. The size of the animals also affects the lift on the plow: taller animals may require a longer chain.

You can also adjust depth by moving the hitch point up or down on the plow. Lifting the hitch point gives the plow less lift, allowing a deeper cut. The lower the hitch point, the more the animals will pull up on the plow, creating a shallower furrow.

If the plow has a wheel, it too must be adjusted. Lifting the wheel allows a deeper cut. Dropping the wheel creates a shallower cut. These adjustments should be made when you first begin work in the field. The plow should ultimately ride easily at one depth and width of cut. If you have to struggle to keep the plow in the furrow, readjust it. A plow adjusted for one type of soil or soil condition may have to be readjusted for another field with different soil conditions.

Adjusting the riding plow is slightly more complicated and requires a greater understanding of the weight of the plow, its wheels, and the weight of the teamster riding on the plow. Ideally, all the weight should be born on the wheels, or plowing will be more work for the animals.

The two-way plow has vertical adjustments that should be made so the plow rides level with little weight on the wheels while plowing. It has a number of adjustments for determining the width of the furrow. The plow may be horizontally adjusted using the position of the plow in the frame, adjusting the tongue-shifting lever, and changing the position of the horizontal clevis.

Every riding plow is different and I refrain from offering but a few words of advice. The simple two-way plow and the single-bottom or sulky plow are appropriate for use with one pair of large oxen. Larger plows with more plowshares or bottoms are more difficult to manage and adjust. They are able to cover more land in one day, but they also require

Horizontal adjustment determines the width of cut; vertical adjustment and chain length determine the depth of cut.

Ancient Roman writing states:

"The length of the furrow shall not exceed 120 paces or else the oxen shall have no time for breath."

OXEN LORE

more than one pair of oxen. Adjusting the riding plow is the same in principle as adjusting the walking plow. Your goal is to maintain the plow so it remains in the soil, cuts a straight furrow, easily turns over the furrow slice, and buries the sod or trash.

Plowing Alone

Many people believe oxen cannot plow with just one person monitoring the plow while driving the team. All over the world and in our own agricultural history single drivers have plowed with oxen using both walking plows and sulky plows.

Give your animals the opportunity to learn and plenty of time to practice. They will soon learn to follow a furrow and respond to commands you give from behind as easily as they follow a path back from the pasture. First, however, the animals must understand exactly what they are to do. Maintain high expectations, offer proper instruction, and be patient.

CONTOURING

Contouring a field aids in reducing soil erosion and also aids in secondary tillage, planting, and harvesting by providing a flatter field to work. Usually before contouring with draft animals, the farmer begins by plowing and loosening the soil. Loosening makes the soil easier to move. Contours may be made over time by allowing the soil to shift downhill naturally, and capturing it by leaving unplowed sections of sod. The sod is left in contours that follow the lay of the land, wrapping around hills and high spots.

Plowing the field in a downhill direction is another way of moving soil. Leaving sod strips or other barriers such as stone terraces to capture the soil will, over time, create the desired contour.

How Much Can an Ox Plow?

Asking how much an ox can plow is like asking how much work a human can do. Dozens of studies from all over the world have tried to answer this question. The amount of work an ox team can accomplish per day is based on how well the team is trained, the teamster's commitment, the soil, the geography, the climatic conditions, and the size of the plow or weight of the implement to be drawn. The animals' size is an important factor, as is the comfort of their equipment. Many people do not allow their animals to realize their potential to work in the field because they do not pay attention to their animals' needs, nor do they have realistic expectations.

An ox needs time to adjust to his work. Research conducted in Africa concluded that even though oxen may lose weight when plowing, they usually increase the amount of work they accomplish after a period of three to four weeks on the task. It seems, therefore, that an ox doing a certain task cannot be evaluated until he has had a few weeks on the job.

Results from numerous studies of the working rates of oxen can be found in *Harnessing and Implements for Animal Traction* by Paul Starkey. While presenting

Plowing with four or six animals reduces the physical exertion required per animal.

a lot of interesting information on the working ox, Starkey cautions that the many variables make it impossible to compare output in different regions and soils or among animals with different levels of training and conditioning. Furthermore, oxen that plow only a few hours each day will be livelier than those that plow all day for weeks on end.

The slowest teams in an experiment reported by Starkey were two animals that managed 0.33 meters per second in Nepal. These 550-pound oxen were pulling a crude ard plow after heavy rains on hilly soils. In contrast, the fastest oxen studied were in India. Oxen weighing 1,600 pounds each managed to make 0.79 meters per second pulling a 6-inch plow in "heavy dark soils."

At first glance these statistics seem intriguing. Neither an ard plow nor a 6-inch plow, however, is something an American farmer would brag about using. The amount an ox can plow is not usually discussed in terms of meters per second. The conditions under which these animals were used are not fully known. Did they plow all day or for a few hours? How were they trained and conditioned? What kind of yoke were they wearing and how might other animals perform under the same conditions? All these factors greatly influence the measured performance of the animals. All of these factors will influence the performance of your animals.

Ox teams in early America often plowed about an acre per average day and could harrow about three to four times that amount. Only animals in top physical shape can be expected to plow 8 to 10 hours a day. Even so, we're not talking about a day of continuous plowing. Unlike anything that is done with a tractor, animals may have their own agenda. They will create rests for themselves by eating, walking slowly when turning and trying to go back to the barn. Animals in poor condition or that are not acclimated to the work will not be able to work a full day.

Oxen benefit greatly from a midday rest to have a couple of hours to feed or ruminate. Frequent water breaks are important during hot weather. Plowing at the rate of one acre per day might be acceptable. Under ideal conditions a pair of oxen might be able to plow slightly more. A multiple hitch could significantly increase the amount that could be plowed in a day, while also easing the burden of each individual animal.

Secondary Tillage

Secondary tillage is used to prepare a seedbed for whatever crop is being planted. It can also aid in weed control. However, it also makes the soil more susceptible to erosion by breaking it into finer particles, necessitating extra time and effort in the field before planting. Many modern systems of no-till or minimum tillage farming stress the importance of conserving soil with minimal disturbance. When using secondary tillage take precautions to ensure that your soil does not wash or blow away between soil preparation and plant growth.

Harrowing

After a field has been plowed it usually must be harrowed, both to provide a smooth seedbed and to level the field. Harrowing creates smoother footing, making subsequent trips around the field easier on the animals.

Many types of harrow are available. One of the most commonly used after plowing is the disc harrow. Single or multiple gangs of discs are placed at opposing angles to cut the soil clods and level the soil. This type of "double disc" harrow works quickly and effectively for leveling a field and preparing the seedbed.

Harrows with two sets of opposing discs, one in front of the other, work better at cutting up the furrow slices and making the soil level. A harrow with a single set of gangs set at opposite angles tends to push the soil into ridges rather than leveling it. Turning the discs creates less resistance than some other harrow designs such as spike-tooth or spring-tooth harrows that are dragged across the soil. The discs are more expensive, but they last longer than metal or wooden spikes and maintenance is easier.

The spike-tooth harrow is used to smooth and level the soil. It does not cut up the clods and move soil as easily as a disc. It penetrates the soil to shallow depths, about two inches, unless a considerable amount of weight is added on the top. Adding weight makes the harrow significantly harder to pull and results in broken or bent tines. The spike-tooth works best for smoothing and leveling a seedbed after plowing in soils that easily break apart or soils that have been harrowed with a disc.

The spring-tooth harrow works by much the same principle as the spike-tooth, but is more effective over rough and stony ground. The teeth penetrate deeper, flex when they hit obstacles like roots or stones, and are less likely to break or stall the harrow.

Make sure the harrow fits your animals. Start with a 4- or 5-foot harrow for small oxen and increase the size as the animals get larger. Spike-tooth harrows may have flexible sections that can be hooked together to make a wider harrow. Expect a pair of animals to harrow at least three to four times as much land as they can plow in a day.

Harrowing requires much less precise control than plowing. The activity may be used to train and condition young animals. Be sure they are not only of an appropriate size, but also of appropriate numbers to pull the harrow. Some large harrows are intended for large numbers of animals or a large tractor. Most harrows designed for use with a tractor will require a forecart. A harrow designed for horses requires little or no retrofitting for oxen.

Cultivating

Many farmers using draft animals cultivate row crops one, two, or more times in order to control weeds. Cultivators work by cutting the weeds at their roots, burying them or tearing them out of the ground. Every cultivator is different. Make sure your cultivator is appropriate for your row spacing, crop, and weed type. During planting, the crops need to be planted in straight rows with adequate space between rows to allow your animals to pass during cultivation. Rows that are crooked and too closely spaced will make cultivation that much more difficult.

Some cultivators travel over the row of plants; others travel between rows. A single animal can pull a small between-the-row cultivator. Larger, more complicated cultivators require multiple animals.

Cultivating with a single ox walking between the rows is a simple system that works well if the ox is large enough for the work. Keep the animal on a halter at first to prevent crop damage.

A disc harrow, designed for smoothing plowed ground prior to planting, makes a superb tool for training a young team.

A horse-drawn over-the-row cultivator with a seat for the driver works well with oxen. Most such cultivators have pedals and levers that offer the operator numerous adjustments and the flexibility to move the tines where they are needed. Over-the-row cultivation is more effective at removing weeds than between-the-row cultivation, because the space between rows is cultivated twice: with each pass, one set of tines travels over the path between the new row and the one just finished.

Unlike harrowing, cultivation requires experienced handlers, well-trained animals, and precise animal control. The hoofs of an ox not following a row, or a cultivator that crosses a row, will damage young plants. The easiest way to control your animals, especially a team unfamiliar with cultivation, is to have a driver beside them and someone behind controlling the cultivator. Oxen that wander, run away, or refuse to pay attention to the teamster should not be used for cultivation. Cultivating is not a recommended activity for initial training of steers.

Using a Muzzle

You may want to muzzle an ox used for cultivation, to help him maintain focus by preventing him from eating the crop. The muzzle may be as simple as a basket made of poultry wire, tied over the ox's muzzle.

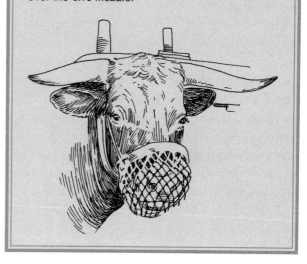

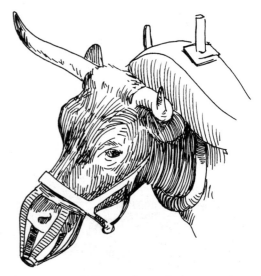

Cultivating between the rows of corn, a single ox wears a muzzle fashioned from leather straps.

Using an over-the-row cultivator requires excellent control of the oxen.

Ridging

Certain crops, such as potatoes, are more easily grown and maintained in ridges. Creating ridges is easily done with a ridger pulled by oxen. The ridger pushes an equal amount of soil to the left and right, forming a trough in the middle and a ridge on each side.

Ridges may also be created with a moldboard plow on soil that has previously been loosened by tillage. Plowing the field in one direction and then reversing the team and plowing back in the opposite direction puts the two furrows together to create a ridge. Have the off animal walk in the first furrow and then turn around and have the nigh animal walk back in the same furrow. The only problem with this system is that it requires the off animal, the one that traditionally walks in the furrow, to be out of the furrow. This request may be confusing for a well-trained plow team.

Seeding

Straight, evenly-spaced rows should be your goal when planting any crop with animals. Straight rows will ease harvesting and are much easier for an animal to follow while cultivating. Crooked rows wear on the patience of both the teamster and the animals. The rows must also be wide enough for the animals to pass easily without damaging plants.

Every crop has a recommended way to plant the seeds. Modern seeders are equipped with plates that can be changed to adjust for seed size and spacing. Seeders can also be adjusted to accommodate the depth the seeds are planted. A seeder designed to be operated with a small tractor may be used with a motorized forecart. Simple approaches are available, although the results may not be as pleasing.

A moldboard plow may be used for simple planting by ox power. In many developing countries, the plow is the only implement. As oxen plow each furrow, seeds for a row crop like corn or beans are tossed into the furrow and the next furrow covers them up. The furrows must be shallow or the seeds will be buried too deeply for germination. The rows will be close together, making weeding difficult.

A system that works well for seeding cover crops, cereal grains, and hay fields or pastures involves first plowing and then harrowing. After harrowing, the seeds are broadcast either by hand or with a broadcast seeder. The oxen then pull the harrow over the field again. With the discs or tines set straight to create little cuts, the seeds are deposited into the soil. This system has its flaws because the seeding depth is not easily adjusted except by adding weight to the harrow. Discs work better than other harrow designs. As the discs turn, the seed, scattered on the ground surface, is deposited into the tiny furrows created by the moving disc. A spike-tooth or spring-tooth harrow may drag the seed across the ground, especially if a lot of trash is in the field.

SEEDERS

Seeding row crops may be done with a seeder designed for use with draft animals. Single-row seeders are relatively simple, but hard to control for straight, evenly-spaced rows. Without a marked field and well-trained animals, a poor job will create a poorly-planted field. Two-row or multiple-row seeders designed for horses are the cheapest and most easily adapted multirow seeders to use with oxen.

A two-row planter requires a team and a teamster who can both walk a straight path.

They maintain straight row spacing between two rows. Most two-row seeders have arms to mark the next row to be planted, helping guide the teamster in animal placement.

The John Deere 999 two-row corn planter is the standard used by most draft animal farmers today. It is easy to use and relatively trouble-free, and its seed plates are readily available. When purchasing your first seeder check to see if the original plates are available. Examine all the working parts to make sure the seeds will be moved from the seed tray or bin to the soil surface.

For planting cereal grains, a grain drill designed for horses is easily adapted to oxen. As for seeders, make sure all the original parts are available and in working order.

ROLLERS

A roller is a great aid in establishing a good stand of grass or cereal grains. Rolling ensures good soil contact by slightly compacting the loose soil to make it less affected by wind or water erosion.

Harvesting Forage Crops

Oxen may be used to harvest many kinds of crops, and most commonly forage crops. Farmers living in temperate climates must harvest forage crops and store them for seasons of little or no plant growth. In northern areas enough forage may need to be stored to last more than six months.

Hay

The most common feed used for oxen is hay, although grains and silage are also fed. These crops offer numerous training opportunities for a team, as well as the chance to employ them on their own behalf. As in cultivating, the ox that wears a muzzle will be easier to work in the hay field. The lure of the grass or newly mown hay is often too much for even the most well-fed ox. Muzzling the animals helps them maintain focus on your cues and the task at hand.

MOWING

Mowing hay with oxen is most often accomplished with the use of a horse-drawn sickle bar mower, although a PTO-driven mower may be used with a motorized forecart. Many early horse-drawn models, including various models made by John Deere, International, and McCormick-Deering, work well

The Driver's Weight

When using a mower that was designed for the driver to ride in the seat, the driver's weight is important to the proper functioning of the mower. If your animals' level of training does not allow you to drive them from behind, have someone else sit in the seat or place a heavy weight, such as a bag of sand, on the seat.

Driving with lines, a single person may mow hay with a sickle-bar mower.

with oxen. Most models were numbered; higher numbers are the more recent models with fewer quirks and problems.

A 5- to 6-foot cutter bar works well with mature oxen. A smaller cutter bar could be used with a mower designed for a single animal.

Finding a mower in working condition can present a challenge. Many horse-drawn mowers have been lawn ornaments or left to rot in the field or farm dump. In evaluating mowers, check the wheel bearings and the pitman bearings for wobble. Open the gearbox and look for gears that are in good shape and still in oil. Gears that are broken or rusted will be difficult or impossible to work with. All the pedals and handles should be present, and the cutter bar repairable.

A mower that has been sitting outdoors for years may take a considerable amount of time and money before it works again. You might consider purchasing a reconditioned mower in an Amish community. Dolly wheels are optional, but worth the extra cost. A reconditioned mower is field-ready and often a bargain for someone who is inexperienced or does not like fussing with equipment.

Most old mowers were manufactured with steel wheels to turn the gears that move the cutter bar. The wheels must have good contact with the ground. The wheels have cleats or tread that help make good contact. If the cleats have been worn off, the wheel may slide and cause the mower to jam in thick grass. Newer models have inflatable rubber tires that not only create a smoother ride, but also may be replaced when additional tread or traction is needed.

Avoiding Heat Stress

Pulling a mower requires a high degree of training and physical conditioning. Since hay is usually mown on hot days, be constantly aware of your animals' reaction to the heat. Oxen that are heat stressed begin to pant. A heavily-panting 2,000-pound ox cannot dissipate enough heat to cool himself on a hot day while pulling a mower.

Unless you can mow on a cool day or in the evening, plan on frequent rests and watering. Oxen do not sweat easily to cool themselves. Pouring or spraying water on them helps cool them down, but offers only temporary relief from excess heat. While an ox may fake being tired by slowing down and seeking shade, do not ignore the heat-stress signs of a heavily panting animal. Too much work on a hot day can kill an ox.

Panting is a sign that oxen are overheating.

Oxen that mow regularly soon learn to follow the edge of the mown hay as a guide.

Mowing hay with oxen requires numerous considerations. The speed of the oxen may influence the mower's effectiveness in a thick stand. A stand of fine hay may be difficult to cut with serrated-edged teeth on the cutter bar.

The weight of the tongue is another concern. Oxen wearing a yoke will bear the weight of the tongue on their necks. The tongue weight of some mowers like the John Deere #4 and the McCormick-Deering #7 wears out the animals faster than designs like the McCormick-Deering #9, which has its gear box behind the axle. Mowers such as the #7 were often fitted with dolly wheels to decrease the burden on the animals and allow more acreage to be cut in a day.

Connecticut ox teamster Ray Ludwig believes that an ox team in good condition using the #7 with regular gears on a relatively cool day can mow about 2 acres. I would argue that they could accomplish more in a day if the mower is outfitted with a set of dolly wheels.

TEDDING

In humid areas hay must be tedded or turned before it is raked and baled. Tedding is done with a hay tedder that either turns the hay over or tosses it up and drops the heavy, greener leaves on top. Many horse-drawn hay tedders are ground-driven. These simple machines require little maintenance or special consideration. They are relatively easy to pull compared to other forage-harvesting equipment. The revolving teeth or fingers do the work.

The tedder fingers should pick up the hay but not strike the ground. The fingers are sometimes broken and must be replaced, but otherwise tedders are easy to use and maintain. A modern gyro-tedder requiring a power take-off may be used with oxen pulling a motorized forecart.

RAKING

Once the hay is dry, it is ready to be raked. Raking with oxen may be done in a number of ways. Using a dump rake — simply dragging it across the field — is one of the older methods. An old dump rake can probably be put into service with minimal time and effort. The rake is lifted every few feet to form lines of hay. Dump, raking is simpler but more tedious than other systems.

Many models of ground-driven side delivery rakes are easily put to use. Older models had three or four wheels: some had one or two wheels to support the rake and two to support the teamster on a seat. These models are the easiest to adapt and use with oxen. Newer rakes manufactured for trac-

With a hay lift and trolley, even a team of young steers may be put to work filling the haymow.

tors, usually having only two or three wheels, may be used with a forecart.

The rake's job is to move hay from its scattered position across the field into neat windrows for the loader or baler to pick up. Raking is an easy job for a team that follows the teamster's directions.

LOADING

For centuries, hay was harvested loose. With the proper equipment and a drive-through barn, harvesting loose hay is easy. Once it is raked or otherwise gathered it may be picked up with either pitchforks or an animal-drawn hay loader pulled behind a wagon. The large wheels power the hay loader; its fingers pick up hay, move it up the loader and dump it into the wagon. Once the loose hay is in the wagon and packed to maximize the load, the loader is unhitched and the wagon is hauled to the barn for unloading.

If the wagon can be driven into the haymow, the hay is then pitched either up or down, depending on the barn design. Farmers in hilly country traditionally built their barns on a hillside to make haying easier. Some barns had animal-powered hay forks and trolleys that picked up large amounts of hay and moved them wherever they were wanted in the barn. Because these trolley systems remained under cover inside barns, many are still functional.

BALING

Hay is baled on most farms today. Many different baler designs offer many options in haying with draft animals. Some farmers do all the hay preparation with animals, but leave baling for a tractor. Most modern balers require a power take-off. Even a motorized forecart must have a substantial motor to run a baler.

Older baler designs frequently used small engines mounted directly on the baler, offering more flexibility as to the type of tractor required to pull the baler. This design offers a way to bale hay easily with oxen. The baler is hitched behind a simple forecart and run by its own small engine.

Oxen can easily pull a baler. The action of the plunger tends to roll the yoke back and forth on their necks, but most animals readily adapt to this.

Raking hay is an easy job for a team that has learned to follow directions; the use of muzzles would keep them from eating up the profits and keep them on task.

A hay loader makes loading loose hay fast and easy.

Loose hay is readily pitched into a hay wagon like this one used on Adams Point in Durham, New Hampshire, about 1930.

A self-propelled baler mounted with its own engine works well with oxen hitched to a basic forecart.

Silage

Harvesting crops that are to be ensiled is more challenging for animals than harvesting dry hay because of the weight of the material. With a good wagon or cart the weight may not be a serious concern, provided the crop is relatively close to the silo. Years ago, farmers hauled green forage to the silo, where it was chopped with a stationary silage cutter. These cutters were slow and probably did not offer much labor savings over dry hay.

Modern machinery for harvesting green crops comes in all shapes and sizes, providing many options. Small round balers will bale green hay that can later be wrapped or placed in airtight bags. Numerous kinds of bale wrappers and movers are designed for use with horses. The power required to process green hay may necessitate a motorized forecart and modern implements, but the labor savings over a stationary silage cutter can be tremendous.

Transport

Oxen have no limits in the ways they may be used for transport. Transport may be as simple as hauling manure to the fields or drawing a cartful of neighborhood children. Oxen have been used to move anything that needed to be moved. Histori-

cally, many examples may be found of oxen pulling everything from cannons to boats across land.

SLEDS

Carts or wagons of all sizes and shapes are easily adapted to oxen. Wheeled transport is easier for them than pulling a sled, but is more costly and challenging to build. Sleds work well in snowy areas, and in warmer climates sleds may be used to give oxen a good workout. Historically most heavy transport was done using sleds in the winter on snow, when wheeled transport was difficult due to the condition of early roads.

A regular part of winter transport is breaking a sled free that is frozen to the snow. Once the sled is moving it goes easily, as the friction of the runners slightly melts the snow. If the sled stops it may freeze to the ground. Cutting the load's weight (pulling the sled sideways just a few inches) is important, and not doing so may discourage even the most well-trained animals.

In northern areas, sleds are used to haul maple sap in the spring. The season starts when the ground is covered with snow; as the snow melts the sled works as well as a cart in the forest where snow remains in shaded areas and mud appears in warmer spots and on south-facing slopes.

A sled is handy for hauling maple sap through alternating patches of snow and mud.

CARTS AND WAGONS

The two-wheeled cart has been the traditional mode of transport where oxen are used. It has remained important in many developing nations around the world. Ox carts are simple in design, cheaper to build and easier to maintain than a four-wheeled cart. This is not to say that it is the best system for using the ox for transport, but for small farmers with limited resources, the ox and the two-wheeled cart are common. The best ox cart is well balanced with only a slight amount of weight on the tongue. The design should not allow the cart to lift the oxen if it is loaded too heavily toward the rear.

The ox cart's versatility is improved with a dump body, which allows easy unloading of manure, gravel, or crops. The dump body is probably the

The two-wheeled ox cart is versatile and maneuverable in tight places.

The four-wheeled wagon works best where large or heavy loads are involved.

most important improvement to the simple ox cart design.

Wagons of many designs offer greater capacity than a two-wheeled cart and also reduce the weight of the tongue, decreasing the burden on the animal. A wagon is harder to maneuver, requiring more room for turns and backing. Any disadvantage, however, is usually outweighed by its capacity to haul larger loads.

Many manufactured designs are appropriate for oxen. For someone interested in welding and cart construction, car axles, tires, and rims are readily available at auto junkyards. The possibilities are numerous for custom-designed ox-powered wheeled transport.

BRAKING SYSTEMS

A major concern is controlling the wagon, cart, or sled going downhill. Since oxen are not usually harnessed using a britchen (breeching), braking a cart is a serious concern, especially if the animals have no horns. Horned oxen wearing a neck yoke will learn to hold a load with their horns. A britchen aids in checking the speed or stopping a load on a hill.

The wagon tongue must be equipped with a stop that will prevent the tongue from sliding through the yoke ring and letting the wagon overtake the team. Severe injuries to both man and beast are possible with even the smallest cart if the oxen cannot hold it back. Make sure the yoke ring is the correct size for the pole and the tongue-stop. A yoke

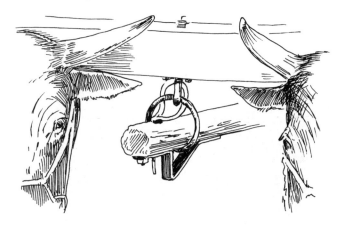

The simplest way to hitch a cart, wagon, or sled to the yoke is with a pin and a tongue stop, which may be made either of steel (above) or from the natural crotch of a tree (right).

This team of cows are fitted with britchen to help them maintain control of the round bale mover when going downhill.

BRAKING MECHANISMS

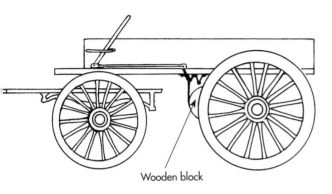

Wooden block

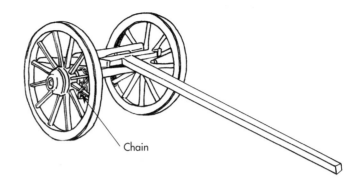

Chain

ring that is too large will allow a small tongue-stop to slide through on a bumpy road where the pole is constantly bouncing. Most wagons designed for horses are equipped with stops, adding to the safety of driving a wagon downhill.

A sled should be equipped with a pole so that even on slight downhill grades the sled does not overtake the animals. If no pole is available, wrapping a chain around the runners will help slow the sled when going downhill and help prevent it from overrunning the animals.

A wagon or cart should have a braking mechanism of some sort. It might be as simple as a chain wrapped around the wheels to prevent them from rolling and overtaking the animals. More innovative ox teamsters design improved brakes that slow the wheeled vehicle and prevent it from overtaking the team.

The size of the load in either a sled or wagon has an effect on how much effort is needed to control its downhill speed. Heavy loads of gravel, logs, or stones must be braked in some manner. Even with britchen or a head yoke, oxen may not be able to control such loads.

Head yokes are traditionally used in many hilly nations and rigidly attach the animals to whatever they are pulling. They have some advantages in controlling a wagon or sled, but problems arise when the load is too heavy or builds up too much momentum for animals that run downhill.

Other Farm Uses for Ox Power

Oxen are a readily available power source that can easily be adapted to a variety of jobs on a farm. In areas with few draft animals, oxen can be used to attract customers to farms trying to capitalize on niche markets and direct on farm sales.

MANURE SPREADERS

Spreading manure is a simple form of transport that uses the animals to haul manure from the barn to the field. For anyone with oxen or other livestock, manure spreading is a never-ending but important task. Manure spreaders not only move the manure but spread it as well, offering great savings in time.

Manure spreaders were originally designed with four wheels and a seat from which a pair of horses could be driven. Such spreaders are simple and easy to use. Many older designs are made of rugged steel with wooden boxes. Replacing the wood and a few broken chain links or gears is all that is needed to put an old manure spreader into use.

The simplest spreaders are ground-driven. Some are intended for use with a small tractor; others are manufactured for draft horses. A two-wheeled manure spreader requires a forecart to carry the weight of the tongue.

FERTILIZER SPREADERS

Many two-wheeled ground-driven chemical fertilizer spreaders have been designed for use with small tractors. These spreaders are easily hitched to a simple forecart, and some may be adapted for use with a pole to be pulled directly by the animals, adding to the versatility of oxen on the farm.

CHEMICAL SPRAYERS

While concern is rising over the use of agricultural chemicals, particularly herbicides and pesticides, they are still widely used in agricultural operations. Draft animals have been used to pull many different types of sprayers and continue to be used today. Some sprayers are equipped with battery-powered pumps, others are driven with motorized forecarts. Sprayers are ground-driven. Although most sprayers are not heavy, the sprayer arms are large and cumbersome. However, most may be folded in normal transport and opened in the field.

The toxicity of the spray is a genuine concern. The safety of both the operator and the animals must be considered. Many farmers using draft animals weigh the pros and cons of using such equipment and choose not to expose themselves, their land, or their animals to potential toxins. Farmers in developing countries and with few affordable alternatives may have no choice but to use their animals to draw sprayers.

Marketing

A team of oxen makes a great marketing tool for a small farm by offering something unique to attract customers. Oxen may be used on a vegetable farm or mixed livestock farm for everything from promotion in a local parade to farm work that is visible to the public. Using rare breeds as oxen can create an awareness of those breeds of cattle.

Oxen can help promote the concept of small sustainable or organic farms. An operation that relies on animal power using homegrown feeds for energy can be a draw for potential clients. Creativity is the key. Enticing people to the farm with ox-drawn hayrides or hauling a Christmas tree out of the forest with oxen are just a couple of the unique ways to draw attention and customers to your agricultural operation.

Using oxen to haul Christmas trees out of a plantation is a nice marketing ploy for attracting customers.

Dick Roosenberg

Scotts, Michigan

When I met Richard "Dick" Roosenberg at Plimoth Plantation, at an ox workshop in 1990, he invited me to come to Tillers International, then in Kalamazoo, Michigan. The following year I took him up on the offer. For the next 11 years, I made an annual trip to Tillers to help run their ox workshops. I was always thrilled to get my hands dirty working oxen in the fields, and I was mesmerized by Tillers's library of draft animal publications. I also made acquaintances with people with similar interests from all parts of the globe, many of whom have become lifelong friends.

While his interests parallel my own, his knowledge and humble nature complemented my skills and ways of approaching challenges.

Every moment spent with Dick was a learning experience. His knowledge of agriculture was extensive and seemed to span the globe. His outreach and experience with audiences both domestic and international were inspirational and truly changed the way I thought about oxen. Instead of having oxen as simply a hobby, Dick introduced me to the importance of these animals to poor farmers overseas right now.

In 1995, I was invited to join a small team of Tillers volunteers to work in Uganda for five weeks. The experience was life changing, opening my eyes to the possibilities of working with farmers, oxen, and livestock overseas. A year later, I was back in Africa, this time in Tanzania planning my Ph.D. research with farmers using oxen. Five years and four trips to East Africa later, I had written a Ph.D. dissertation on *Maasai Oxen, Agriculture and Land Use Change.* As I write these words,

I am planning a year-long trip to Africa to teach at a university in Namibia. Looking back, it is crystal-clear that Dick Roosenberg changed the course of my life.

Dick has been the director of Tillers International for 25 years. Tireless in his work, he has truly achieved the impossible: making a career running a nonprofit educational site that has engaged thousands of people in the use of historic technology in rural development and historic preservation. He has the ability to be optimistic when there seems little to be optimistic about. He engages and works successfully with people at all levels, from fundraising in rotary clubs and boardrooms to teaching vocational students in Africa and farmers in Nicaragua.

Dick is methodical and careful in his planning and thoughtful in his preparations and words. I always found him to be easy to work with.

As an example, Dick let me loose in Uganda in 1995 just to go out in the field and work with farmers and their oxen. As I look back, I realize how naïve I was and how little I knew about how agricultural development worked. I am amazed he did not pull me aside and give me a clear directive about what I ought to be doing. Instead, Dick let me find my own way, while in his own subtle manner he shaped my thinking, my teaching, and my efforts in the field with farmers.

Dick has helped train many budding ox teamsters and offered many experienced ox teamsters like myself new ways of looking at problems. He has provided opportunities for me and countless others to work with animals and historic implements in the field, at Tillers International and abroad, while never telling us what to do, but instead offering encouraging words and helpful hints. He is truly a man to be admired for his vision, for encouraging others to try combining the new with the old, and for his vast knowledge and interest in oxen and agriculture.

This two-wheeled ground-driven manure spreader
requires a forecart when used with draft animals.

A custom-designed lawn mower is used for training a team
of calves while accomplishing an all-too-common farm task.

If you use a sprayer, try to create as safe a system as possible for both yourself and your oxen. Use chemicals with low or no surface toxicity and a sprayer that is designed to spray low to the ground to reduce the chance of drift and contamination.

STATIONARY POWER

Long before internal combustion engines, oxen were used to power many different machines that required turning a shaft or gears. The animals were hitched to a pole or sweep and walked in a circle to power stationary threshers, grinders, water pumps, and even hay balers and bucksaws. Today such machines are most often seen in farm museums, but animal power in other parts of the world is still used for pumping water and grinding grains.

Another stationary power source came from treadmills that required animals to walk on a floor that moved and thereby turned gears. The animals were usually enticed into walking by putting feed in front of them. Treadmills were used to create power to run everything from sawmills to butter churns and washing machines.

CHAPTER ROUND-UP

Despite increased interest in draft animals in the United States, ox power has not been seriously used in agriculture for decades due to a lack of understanding of training and using cattle. As the most important draft animal in the world today, however, and the beast that paved the way for other draft animals in this country, the ox certainly has a place in modern agriculture.

Farming offers many possibilities for using oxen. Horse-drawn implements are more available and often cheaper than tractor implements. Combining ox power with modern equipment, perhaps by means of a forecart with an attached internal combustion engine, presents exciting possibilities.

Oxen offer flexibility and readily available animal power. The key is to use them appropriately. Work them regularly, employ them in as many tasks as possible, expose them to different environments, and you will have well-trained oxen willing to work long hours at plowing and other heavy work. ★

12
·≫ LOGGING ≪·
— WITH —
OXEN

Few activities a teamster could devise will train oxen better than logging will, provided the animals understand the basic commands and are started with light loads. The team learns to avoid obstacles and draw loads of different sizes under various conditions. The repetition of moving log after log, load after load, is a wonderful teaching tool for young animals.

Serious logging requires well-trained animals that have been gradually conditioned to the work. Although horses and mules are stronger and faster, particularly in warm weather, a team of oxen is hard to beat when it comes to simple logging with a minimal investment of money and training time. For a part-time farmer or the owner of a small woodlot who needs to move firewood or thin a stand of trees, logging with oxen is appropriate, cost-effective, and fun.

Oxen in Logging Camps

Author Robert Pike, in *Tall Trees, Tough Men*, describes both the ox and the logging camps at the turn of the twentieth century in New England: "Before the days of the horses, a team of three yoke of oxen, hitched in tandem, was ordinarily used to haul logs to the river. It took a skillful teamster about two months to break a green team."

The author claimed that oxen don't get sick as often as horses do, "But they would get lame and galled [stabbed by another ox or by dropping trees], and when the snow was crusty they were always cutting their legs below the dew claws." Cut legs required frequent treatment. Oxen numbers dwindled by 1890 because of the difficulties with shoeing, their slow speed, and their inability to slow a load adequately. At that point, "all but the most stubborn New England ox men made the transition to horses."

According to Ralph Andrews in *This Was Logging*, oxen were worked in the Northwest into this century. Although they were often referred to as "bulls," the animals' physical characteristics, shown in the early photographs in his book, make it obvious that many were castrated mature bulls. Most of the early cattle were nondescript animals with characteristics of the Shorthorn and Devon breeds.

To move the huge logs of the Pacific Northwest, numerous teams had to be yoked together. Twelve oxen (six pairs) were frequently hitched to logs that were pulled on greased skid roads.

The life of a "bull" in the logging camps was not easy. Andrews reports that "Bulls were kept in corrals and barns at night, [and] shod with two plates to each cloven hoof on Sundays and days when the weather or snow was too bad to work."

Author Dave James, as quoted in *This Was Logging*, wrote, "The oxen were brought in from great distances, chained and yoked together, [and] goaded into effort by bellowing men that could be heard for miles. Boys . . . were given the job of swabbing whale oil on the skids to help the logs slip. This was about the only help anyone gave the bulls. The [bull] puncher, that master of profanity, consigned them to burning eternity a hundred times daily.

"Of all the types of power brought into the timber, only the bull teams served a man's stomach. The accident victims could be eaten."

Historical Practices

Oxen in the United States were used for logging in the Northwest, the South, the upper Midwest, and New England, where logging was usually a winter activity. Cattle cleared the New England forests of large virgin timber in the 1600s and 1700s, and in many areas continued to be used in the woods well into the 1900s. With little outdoor farm work, many colonial farmers became lumbermen during the winter, their farm teams transformed into woods teams.

Freezing temperatures, frozen waterways, and snow or ice benefited oxen in many ways. Provided the team was shod, they could draw tremendous loads over rough ground or swamps that would otherwise be impassable. Other advantages of winter work included a lack of troublesome insects and the opportunity to increase the animals' stamina without concern about overheating the team.

For centuries the ox team was the chief power source for logging in the United States. As the lumber industry became more commercialized they were gradually replaced, nationwide, by horses. During this transition the two animals were used in combination. The oxen primarily skidded logs to landings. Once the logs made it to the landing, they would be loaded onto huge sleds pulled by horses and in later years loaded onto trains.

Oxen were slow, steady, less valuable than horses, and well suited to the difficult and dangerous work of logging. Their pace always offered plenty of opportunities for coaxing and goading by the teamsters. In dangerous areas they may have been maintained and used to prevent injury to horses. Even when an ox was injured, the loss was not total because the animal could be used for beef.

Horses largely replaced oxen by the end of the nineteenth century in American logging camps, and machinery later replaced both oxen and horses. While draft animals still have advantages over modern machinery, particularly in small lots or environmentally sensitive areas, oxen will never compete in the woods based on volume or efficiency.

Logging and Pulling

Although few teamsters in New England today log commercially with oxen, many continue to work in the lumber industry. Most use contemporary methods and machinery on the job and take advantage of being self-employed to take the necessary time off to work with and condition their teams for competition.

Pulling contests in New England occur during the summer and fall. The presence of oxen is a reminder of the early days when teamsters came from the forest in spring with teams well conditioned and "hard." They challenged friends and neighbors to test the strength of their teams and began New England's long-standing tradition of ox pulling.

In New England, most logging with oxen was done in the winter, when the farmer and his team had spare time and large loads could be hauled on sleds over snow or ice.

During summer months, a cart or wagon was used to haul firewood or pulpwood that had to be moved any appreciable distance.

Ox Logging Today

Many current New England ox teamsters log part-time, using oxen for small-scale forest thinning operations or to haul their own firewood. Oxen are still commonly used for logging in Latin America and Africa. Although they do offer some advantages over machinery, their slow pace detracts from their ability to compete against modern machinery, where it is available, and even against horses.

Compared to horses and mules, oxen may be slower, but they are less likely to shy from chain-

In New England, logging is often a 4-H event that creates a fun learning environment while helping train teams and raise funds for other projects.

Hauling small logs, like these at the New Hampshire Farm Museum, makes a good training aid for young steers.

saws, heavy machinery, falling trees, and deep snow or mud. During cooler winter months they compare favorably to horses and mules. In summer their pace will become slower than usual, because of their inability to sweat and thereby to dissipate body heat.

Unlike with horses, the ox teamster leads from the front, a vantage point that offers safety in avoiding the many obstacles such as mud, rocks, stumps, and other logs. The teamster is ahead of anything the team is pulling and therefore has no need to dodge a twisting, rolling log.

Good woods horses don't need a driver because they can be directed to take a log to the landing and return to the woodlot for the next log. Oxen can meet the same challenge, but only partway. A well-trained team will readily pull logs out to a landing, but I've never seen a team go back into the woods by themselves out of sight of the teamster.

Animals as young as six months old may be used in the woods, but for only a few hours at a time hauling small-diameter 8- to 12-foot poles for firewood. Yearlings may be expected to work most of an eight-hour day provided they get appropriate breaks and plenty of feed and water. A more mature team may be expected to work a full eight- to ten-hour day once they are used to the work.

APPROPRIATE WOODLOTS

When logging with oxen, first ascertain that the woodlot is appropriate for animals. The layout and lengths of the skid roads are critical. Steep, rocky slopes are challenging, and wet swampy ground should be avoided unless it is frozen. Mud dirties the logs and quickly tires the animals.

Slight downhill skids of only a few hundred feet are best, especially if you are skidding individual logs. Longer skids or uphill skids will wear out an animal faster, particularly if the weather is warm or the ox is worked only occasionally.

Keep each animal's condition in mind. Nothing is more discouraging to a team of oxen than being pushed beyond their psychological or physical limits. Be careful not to overwork a team that is unaccustomed to the activity.

Logging in snow is a good way to teach a team to drive from behind, as they are not eager to leave the packed trail and enter deep snow. Logging on dry ground is difficult work, especially for a young team. On long hauls the team will benefit from frequent rests.

GROUND SKIDDING

The basic requirements for working in the woods are a yoke, a team willing to work, and a chain about 12 to 15 feet long, with a slip hook on one end and a grab hook on the other. The end of the chain with the slip hook is wrapped around the log or logs and the grab hook is used to attach the other end of the chain to the yoke, allowing adjustments for length. Many teamsters cover part of the chain with plastic pipe or hose to protect their animals' legs from chafing when they must turn a corner with a log.

Ground skidding with a chain works best for twitching logs short distances. The most common scenario is to pull the logs to a common area, or landing, where they are loaded onto a wagon or truck. Minimizing the distance the logs must be twitched on the ground keeps them cleaner and is easier on the animals.

A few precautions must be taken when twitching logs. Logging was historically done in winter to make twitching and hauling easier for the animals. During warm weather the forest floor has exposed roots, rocks, or stumps the team might get hung up on. A log that catches on an obstacle will slow the oxen or even stop them in their tracks. If the team has not learned how to move from side to side to free the load and instead they continue trying to pull, they may break the yoke or chain.

Logging Implements

Using a chain to skid logs along the ground has its limitations, especially if you have a long distance to travel or the trail or woodlot is particularly rough. Be aware that some sawmills won't accept lumber that has been skidded through dirt because of the chance of damage to saw blades.

When ground skidding is inappropriate, a few techniques may be employed using implements that require a greater investment in equipment and larger skid roads. Since initially approaching and hitching these implements to one or more felled logs may be difficult, they are usually used in conjunction with ground skidding.

Most New England ox teamsters use cattle in the woods to haul firewood to heat their homes.

BOBSLEDS AND SCOOTS

In winter a bobsled may be used to raise the front end of the logs to prevent hang-ups. A bobsled is similar to a logging lizard but is used for more than one log. It is designed only to lift the ends of the logs, and not to carry logs piled on it.

Alternatively, a scoot lets you carry the logs or firewood completely off the ground. A scoot is like a bobsled, but larger, usually 6 to 10 feet long, and is used for hauling small logs such as cordwood or pulpwood cut to 4-foot lengths.

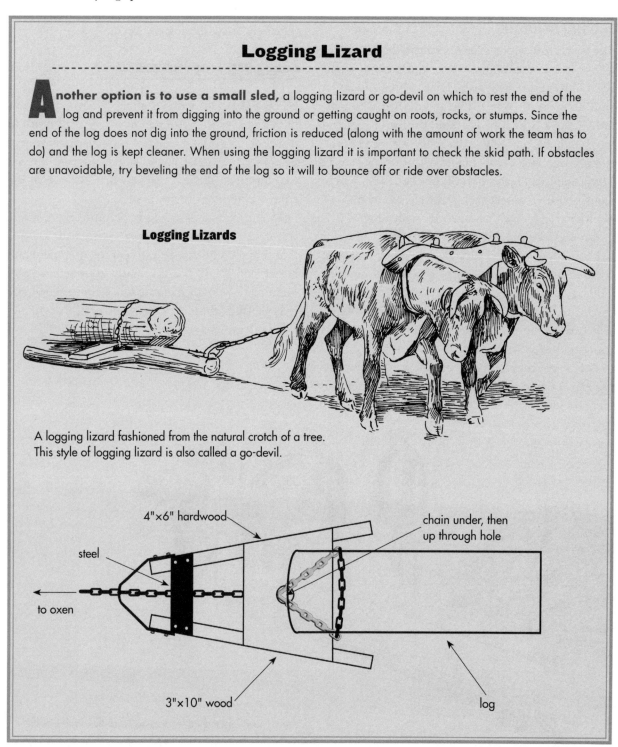

Logging Lizard

Another option is to use a small sled, a logging lizard or go-devil on which to rest the end of the log and prevent it from digging into the ground or getting caught on roots, rocks, or stumps. Since the end of the log does not dig into the ground, friction is reduced (along with the amount of work the team has to do) and the log is kept cleaner. When using the logging lizard it is important to check the skid path. If obstacles are unavoidable, try beveling the end of the log so it will to bounce off or ride over obstacles.

Logging Lizards

A logging lizard fashioned from the natural crotch of a tree. This style of logging lizard is also called a go-devil.

4"x6" hardwood

chain under, then up through hole

steel

to oxen

3"x10" wood

log

Winter Traction

Using a bobsled or scoot over frozen ground or a well-packed trail works well if your oxen are shod. If the animals are not shod, working them in winter may be challenging because their cloven hooves offer little traction on a packed and frozen trail.

Be especially careful when going downhill. Make sure the bobsled or scoot has a pole so the load doesn't creep up on the animals and overrun them. The team should also be accustomed or trained to holding back a load. Oxen that are not equipped with a breeching must learn to use their heads and horns to hold back a load.

The New England logging scoot, usually outfitted with a pole to increase control on hilly terrain, works well for hauling small logs and firewood in snow.

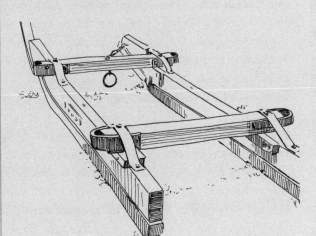

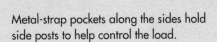

Metal-strap pockets along the sides hold side posts to help control the load.

At the landing the side posts are removed and the logs are rolled onto a pile, to be sawn or picked up by a log truck.

FORECARTS

Many teamsters in the United States historically used carts with large wooden wheels to straddle the log and lift one or both ends into the air, making the moving of logs over great distances or soft ground much easier. In warm weather or southern climates today, wheeled forecarts are used to raise the front of the log off the ground.

Some forecarts use a small winch to raise the log. Others have an axle that drops when the cart is backed up and rises when the cart is pulled forward, raising the front end of the log. This system requires hitching close to the cart.

A kind of arch commonly called big wheels, timber wheels, or logging wheels was once used In flat regions.

Using a small logging arch, this Milking Shorthorn team easily handles this small maneuverable logging aid.

In warmer months, a wagon or cart may be used to haul firewood out of a woodlot. A simple but rugged cart may be welded together from a truck axle. It should be strong enough to withstand hitting an occasional tree, rock, or stump without falling apart. Such a cart may easily be designed to hold one-third to one cord of wood in 4-foot lengths.

Make sure the load is well balanced. Too much weight in the front will quickly tire your oxen. Too much weight on the back may lift the yoke off their necks. Depending on the unevenness of the trail, you may need a chain and binder to hold the load in place.

Making Money

Logging with oxen requires creativity if you wish to generate enough income to cover the cost of your time, equipment, and animal upkeep. Many ways have been devised to charge for the animals' services. An option for small or difficult woodlots is to charge by the hour. Other options include being paid by the day, by the load, or by the amount of lumber or firewood hauled out.

Cleaning up a woodlot that was harvested with machines can generate income with less work. Hauling out hardwood treetops, usually left after the logs are taken, can generate considerable income from firewood without the need to fell trees. Such wood can often be taken for free by arrangement with landowners who want their lots and trails cleaned up. Selling the firewood at the landing will minimize your labor requirement.

CHAPTER ROUND-UP

I grew up logging with oxen. In high school I made a lot of money logging on weekends and during vacations, and the time I spent in the woodlot with my team seemed to whiz by. More recently I have used oxen to harvest firewood and building timbers from the woodlot behind my farm. I have enjoyed few activities in my life more than working with a team in the woods. ★

Brandt Ainsworth
Franklinville, New York

Brandt Ainsworth can't remember exactly when he began logging, because he started as a boy, helping his dad in the woods in western New York near his hometown of Franklinville.

He did start logging professionally on his own in 1995, using horses, and began using oxen in 1999. "I watched the ox show and pull at the Warren County Fair in Pennsylvania, and met Howard VanOrd, and then I guess I got the fever to have a pair of oxen."

I first met Brandt at Tillers International in 2002. In 2006, he helped Tim Huppe and me with a workshop on logging with oxen and was a wonderful asset to the program.

Brandt has written articles and produced, with *Rural Heritage,* one of the best videos on logging with draft animals that can be found. His first pair of oxen, featured in the video, were Limousin-Holstein crosses.

"They were a mistake," Brandt says, "as a beef bull got into a neighbor's pasture with some Holstein heifers, and my first team started with an investment of $60."

When the team, named Timber and Jack, were about six months old, Brandt took them into the woods because they fit nicely in the truck with his horses. "I realized they could pull small logs, and they started going with me daily after that. They couldn't really work hard until they were about 18 months old, and by two years old, they could finally pull some good-sized logs."

Comparing horses to oxen in the woods, Brandt admits he has been more of a horse person his whole life; he guesses he has had 50 horses, most of them bred on his family farm. He has pulled, shown, and logged with horses.

"Different situations work better for each animal," he says. "Horses are quicker and they are a lot better on a long skid, especially if I use a logging cart. However, for short hauls and ground work, the oxen have some advantages. With horses hitched to a log on the ground, you have to work behind the animals, near the log, which is dangerous. You also need more equipment for horses and have more breakdowns.

"The advantages are oxen have is the simplicity of the equipment, fast hooking and unhooking, and ease of handling, especially through tough spots. The oxen can pretty much go anywhere I can walk." Brandt had only one breakdown with oxen, when he hitched the team to a large log high on a bank above the skid road. When the oxen pushed into the yoke, the bow snapped, because the line of draft was too high. Even a good team can have a bad day."

Brandt sets training goals for his teams and work toward those goals. He finds the more real work he has for the team, the better they get. He not only logs with his oxen, but has also taken them in parades, farm field days, fairs, and even pulling contests. "Be persistent and stick with it, he says.

In terms of making money logging with draft horses or oxen, Brandt, a very careful and methodical logger, offers this advice. "You need to constantly improve your logging skills, especially in terms of buying timber and marketing logs, if you want to make money. Don't worry about production: you seem to get more out, and do a better job, if you work a good day and do not set goals and rush to get, say, 60 logs in a day."

Another advantage Brandt sees with oxen is that he would not regularly work a pair of horses in the woods if they were not shod. In contrast, he never shod his oxen once in the four years he worked them regularly in the woods, finding that their feet really "hold up better than the horses."

Brandt recalls, "I had Timber and Jack for five years, and they did farming and logging almost daily." Having trained four other teams, he admits, "If I was only doing firewood or logging part-time I would prefer steers. A team of oxen are cheaper to keep, and I really enjoy oxen."

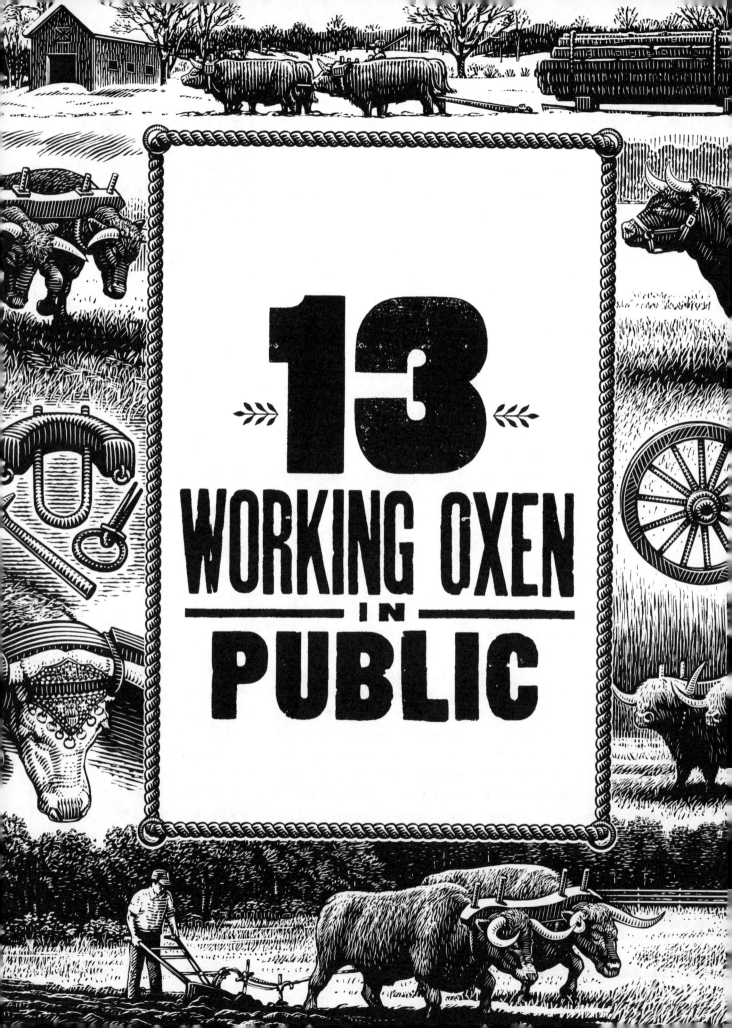

13
WORKING OXEN
IN
PUBLIC

In some parts of the world, oxen are as common as cars are in America. The animals become acclimated to walking long distances and meeting automobiles, trucks, and other animals on the road. In the United States and Canada, most ox teamsters keep their animals for fun and hobbies. The oxen are not usually required to work in the midst of traffic and pedestrians. Yet if they will be taken out in public for any reason, they must be prepared for what they might encounter.

Many ox teamsters have opportunities or invitations to exhibit their animals in public. Holiday parades are popular, and many summer fairs and field days welcome teams of oxen. Other opportunities exist with elementary schools studying local history, living history farms, and competitions in pulls and shows. Oxen adapt well to public situations because they do not easily spook.

Working oxen in public can be great fun because the lumbering beasts are usually calm and easy to work with. As a boy of nine or ten, I vividly remember watching a teamster drive his pair of giant Holsteins using only voice commands and a small stick. From that day forward I was obsessed with having a team.

. . . .

Oxen adapt well to public situations because they do not easily spook.

. . . .

While the experiences I have had with oxen in public have all been positive, I have seen many teamsters who were not so lucky. Preparing for the unexpected is the key to safe public appearances.

Training Level

Well-trained oxen follow their teamster, watch the teamster carefully, and listen for commands. The animals have to learn to pay attention in all situations. Traffic, sirens, horses, or other cattle can scare a team. An important part of training is exposing your animals to such distractions, which means training them to accept new situations, loud noises, and crowds of people.

Oxen in much of the world wait patiently when given a break from work. Work is as much a part of their daily routine as are eating and drinking. In comparison, oxen in the United States these days

Oxen on the road in Tanzania learn to avoid pedestrian, vehicular, and bicycle traffic with little direction from their teamster.

often work only a few hours per week, which means the animals have to be trained to stand and wait.

Before taking your oxen into the public make sure they will stop without hesitation and will turn when asked. They must learn to stand patiently.

A team shown in public must be trained not to fear loud noises like blaring horns and sirens.

Untrained calves should not be brought before the public unless they are securely haltered and under the control of someone who can handle them if they panic.

MAINTAINING CONTROL

Distractions you expose your team to should include chainsaws, heavy equipment, other animals, and large groups of children. In parades or other public events where machinery and sirens may be encountered, you must have a team you can control. When I take a young team into the public, I keep a rope on the nigh steer so I can grab it if necessary. The same should be done with a larger team that has spent a limited amount of time in the yoke.

> Cattle will not forget a painful or disagreeable experience.

Well-trained oxen are exposed to a wider variety of situations than is the typical bovine. Many dairy or beef cattle will shy from a freshly turned furrow on their way to the pasture. Cattle that have been trained to work should follow their teamster without hesitation, even into areas that might spook other animals.

Following the teamster without question requires a level of trust in humans that goes beyond what most cattle have. This trust and understanding between oxen and teamster come only after many hours of work in a variety of situations.

Cattle will not forget a painful or disagreeable experience. If they have been led to an area where they have fallen down, slipped, or been frightened, they will remember the experience for a long time. Training never stops and a good teamster never leads a team into a situation that the animals are not ready for.

ANIMAL AGE

Calves are easily and safely acclimated to distractions. With older animals, training becomes more challenging. Animals that have not been exposed to large groups of people or the sights and sounds of parades and agricultural fairs until they are mature are more likely to be scared. Large oxen can be dangerous when frightened, because they are difficult to control if they try to run or jump away from the situation.

Training oxen as calves, as well as getting them accustomed to the public at a young age, is the best way to bring up animals that can deal with a variety of situations. Young teams of steers, however, cannot work long days or bear the environmental conditions that are involved in a long parade or competition.

Calves are also more susceptible to diseases than are older oxen. The stress of trucking, working, and exposure to new things, including unhealthy animals, can all cause problems while your calves are still developing their immune systems. If you want to use a team of oxen in public, plan your purchase and training program so the animals will be at least six months old, and preferably twelve, before they appear in public for the first time.

Teamster Skills

Teamster skills are a result of driving your oxen and the experiences you share with them. You can also learn much from other teamsters, whether they are experts or novices, through good observation and asking questions. A teamster never stops learning while training and working oxen. Whether it is a new trick or some training aid, never be too stubborn to

Ready for Anything

Work your oxen amid the worst distractions you can create at home, and they will have fewer problems when out in public. I have had my share of mishaps at home. I once had a team turn a corner too sharply, or misunderstand my command, and tear off a wall of an outbuilding.

The only time a team truly got out of my control was also at home, while pulling a stoneboat. As I was walked beside the team I heard a loud hissing noise but could not identify its source. The sound got louder, persisted for a few seconds, and then stopped for a minute or more. My team paid it little attention.

A few minutes later a hot-air balloon basket scraped the tops of the trees just over our heads. The couple in the balloon cheerily yelled hello with a big wave and then released some gas to heat the air and coax the balloon away from the treetops.

The combination of the loud hiss, the shouting people, the basket scraping the treetops, and the giant colored balloon was too much for my team, which took off as if they had lost their minds. It was the first time I felt out of control while working an otherwise well-trained team. I was just lucky it happened at home.

A team that is taken out into the public must learn to stand and wait patiently.

Working on a movie set involves long waits, long days, and lots of strangers.

Riding Oxen

Sooner or later, many ox teamsters try riding their oxen. This activity can be impressive to bystanders, but make sure you maintain good control, and do not invite others to join you. Riding an ox is much more dangerous if the animal is pulling a vehicle or implement. A fall from a tall ox is one thing; falling and then getting run over by a scoot or wagon is an altogether different experience.

A well-trained team of oxen will astonish a crowd. A poorly-trained team or a teamster who is eager to use the whip will sour most people on driving oxen. Do your training long before the day of exhibition. If your team acts up or becomes frightened, blame yourself for not providing them with the proper experience. The best teamsters can anticipate what their teams will do before the animals act.

EXPECT THE UNEXPECTED

When appearing with your team in public, do not lose sight of the fact that the animals have their own personalities and agendas. Different animals react in their own unique ways to various stimuli.

Be prepared for small children to appear suddenly, run at your animals, and grab them in the most unexpected places. I've had children climb on the backs of my resting oxen or grab an unsuspecting ox by the sheath hairs in the middle of its belly.

I once had a pair of yearlings that hated small children. It took me some time to figure out why they disliked kids. Then one day I caught my three-year-old brother in the barn whipping my oxen as they stood in their stanchions. You cannot be with your team all the time. You can never know all that a team has seen or been exposed to. You never know how a team will react until they are put in a situation. But you can get to know their idiosyncrasies and how to control them. Learn to make wise choices as a teamster.

learn or teach your animals something new. A good teamster knows when the team is ready for exhibition. Do not exhibit a team that is too green for public display.

When oxen appear in public, both the team and the teamster are on exhibit. How people view your team is a function of the time and effort you put into their training. Our history books note the ox teamster as a ruffian whose "skills" included whipping and cussing. This method of handling animals is not recommended, and particularly if the animals are in the public's eye. A good teamster needs no more than a slight movement of the whip and quiet verbal cues to control the team. Do a good job training your team, because you may be the only ox teamster ever seen by the people you encounter.

You can impress the public in any number of ways by showing what your team can do, such as pulling a car in a parade.

EDUCATING THE PUBLIC

When out with your team among the general public, do not leave your animals unattended. People will tease them, try to grab their horns, and attempt to pet them without asking your permission beforehand. Make it a practice to let no one touch your team without your permission. If people ask to pat your team, let them do so in a controlled manner, preferably one person at a time.

Sometimes large groups of people, especially children, will run in from all directions to grab and pat a team. They don't know enough to respect the animals, their tremendous strength, and the dangers of being stepped on, kicked, or poked with a horn. Because oxen are slow and generally calm, people forget that they are large and can cause injury just by turning their heads or stepping on a toe. The public must be educated.

Whenever you show your team in public, be ready to answer a barrage of questions about them. Most people think of oxen as just huge beasts they have read about or seen in books. They have no idea what kind of animal an ox is, let alone the amount of time that goes into its training. Being prepared to answer questions about your team is the best form of public relations an ox could have.

Know Your Event

Be sure your team is well adapted to the kind of exhibition you attempt. Cattle can easily handle

---

Weather Considerations

Many exhibitions occur during warm weather, when temperatures make it comfortable for humans to be outdoors. A fine temperature range for working cattle is 50 to 60°F. Be wary in warmer weather if your cattle are not adapted to such conditions. Cattle, especially *Bos taurus* or European breeds, do not tolerate heat as well as horses, mules, or Zebu-type cattle.

Frequent watering and hosing off will greatly aid the comfort level of your team, especially in high temperatures and humidity. Cool weather is more appropriate and comfortable for most working cattle. Be particularly careful with calves, because they easily become overheated and dehydrated.

---

parades, which are often less than a mile or two long. Wagon trains or all-day exhibitions are much more exhausting. Hobby teams that are worked only a few hours each week need to be conditioned before attempting such an event.

RATE OF TRAVEL

I was once in a wagon train that was to cover about 20 miles a day. The event looked fun and exciting. My three-year-old Dutch Belts were fast on their

Sharing your knowledge about oxen is both fun and rewarding.

Grooming

Any animal presented to the public should be in good physical condition and well-groomed. No one is impressed with cattle that have urine stains on their bellies and legs and manure caked to their hair. Most cattle will stay relatively clean if kept outdoors in a clean lot or pasture.

Oxen that are kept indoors, especially in stalls designed for cows, quickly become stained and dirty because they urinate under themselves, and they lie down more than other draft animals. Before showing them in public you may need to wash your animals. Cattle that are accustomed to being washed learn to enjoy it, especially on hot days, provided you don't squirt water or soap into their eyes or ears.

A well-groomed team with hair clipped, brushed, and washed, horns sanded and polished, fascinates the general public. To add to the impressiveness of a team, many teamsters add brass horn knobs, bells, and a masterfully carved yoke and whip. These all portray a wonderful image of the animals and the teamster.

Washing steers or oxen makes them look their best for a parade, show, pull, or other public event.

---

feet and full of energy. In an eight- or ten-hour day the planned speed seemed appropriate for my team and me. I had never been on a wagon train with horses and pony teams, but I soon found out that we weren't going to travel at 2 to 3 miles per hour, the comfortable speed for oxen. In the first two and a half hours the wagon train traveled 12 miles, including two 15-minute breaks. The horse teams were traveling at about 6 miles per hour.

My oxen kept up for about 6 or 7 miles, but I had to trot them down hills. After the first hour it was obvious that my team couldn't maintain that pace. My first wagon train ended at noon on the first day. My animals were not prepared for traveling at such speeds, and the hot weather complicated the stress they faced going up and down the mountains.

POSITIONING THE TEAM

In lining up for a wagon train, parade, or other public event, the last position is usually best because oxen walk slower than most other animals or humans. Avoid walking near a band or fire department because the loud unexpected noises they generate can spook even a well-trained team. On the other hand, oxen pulling a cart or just walking in a yoke often spook horses. Since oxen adapt more readily than other animals, be respectful of other participants.

THE TEAM'S PRIOR EXPERIENCE

Be sure you and your team have worked with the implement or wagon you plan to pull in a public event. Maneuvering a large wagon or cart takes

Bill Speiden
Somerset, Virginia

Beginning with a bet that he could train a steer to be ridden, retired dairy farmer Bill Speiden started working with oxen in 1978. "Art Hine stopped by the farm on one of his trips to Williamsburg with a team," Bill recalls. "He got me straight on how to make a yoke and a number of training tips. I bought my first-off farm-made yoke from him and used it as my prototype for additional yokes since then."

I first met Bill at an ox training workshop I was helping run in Missouri in 1995. His genuine interest in oxen and history was obvious. He asked more questions than most participants and really wanted to improve his training and driving techniques.

His oxen have taken him as far as the Bozeman Trail in Montana and Wyoming, where his team pulled a wagon for 300 miles, and to Idaho, where he worked for the Oregon Trail Center. These days he typically attends 10 to 14 events a year closer to his home in Somerset, Virginia. In addition, however, he loads up his teams for the annual trek to Tillers International in Michigan for the Midwest Ox Drover's Gathering.

"I feel that public exposure to oxen and their history is an important contribution to people's grasp of the real world of the past," Bill says. "Being kind of an adventurer, I really enjoyed the wagon trains, giving me and my helpers an opportunity to experience taking the oxen over 300 miles on an outing, and getting as much as possible of a feel of what

it was like to travel that way in the 1700s and 1800s."

Bill has worked with many breeds, including the Holstein, Brown Swiss, Ayrshire, Milking Shorthorns, Dutch Belted, and an all-black team that was a cross between the Brown Swiss and Holstein.

He says, "The Holsteins worked well in crowds for me, and were easy to teach to plow. Pure Brown Swiss and Shorthorns can be laid-back to a fault. The Holstein–Brown Swiss cross worked well for me, but they tended to get up over 3,000 pounds each, and that 75-pound yoke kept getting heavier every year. I have settled on the Milking Shorthorn–Holstein cross as my favorite. They rarely get much over a ton each, and the red or blue roan coloring are historically correct for early America."

Bill's most memorable moments have included: "All the local people in Montana and Wyoming coming out to see the oxen at campsites along the Bozeman Trail." He also said he enjoyed taking the oxen and wagons through streams and up the

banks of rivers, where the horses and mules could not go.

Watching Bill's slide presentation at the Midwest Ox Drover's Gathering in 2002, a few things stuck out in my mind. First, Bill had to shoe his oxen when they lost shoes on the trail. Throwing a 3,000-pound ox down and shoeing him tied up on the ground was no easy task, possibly resulting in the animal breaking someone's ankle while tossing his head around. To hear Bill tell the story of his time on the trail, you know it was a lot more than just working oxen in public.

Bill's knowledge of the history of the trail and the use of oxen were second to none. There is no doubt that he truly experienced a little part of what he regularly tries to recreate.

Take extra precautions when giving rides in your ox cart.

some getting used to by both the teamster and the team. If you will be traveling up and down hills with a large wagon or cart, your team must be accustomed to holding back a load. The dangerous combination of a heavy wagon, no braking device, and a team without shoes or experience in holding a load can lead quickly to disaster.

Insurance

When working large animals in public, invest in some type of insurance with a liability policy on your animals in case they do run away or get out of control and cause property damage or injury. Many farm/homeowner insurance policies can include such coverage. Insurance is usually cheaper and more readily available if you assure the insurance company that the general public will be only viewing your animals and will not be invited to ride in your ox cart or wagon.

If you do offer rides, take extra precautions. Although runaways are not likely with a well-trained team, there's always the chance your animals will get spooked and jeopardize the passengers' safety.

CHAPTER ROUND-UP

Exhibiting oxen in parades and competitions and at field days and living history farms or schools can be a wonderful experience. The presence of a well-trained ox team is often the highlight of such events. Oxen adapt well to being in the public eye because they do not spook as easily as other livestock.

Although sharing your experiences with others and impressing a crowd with your well-trained team is fun and rewarding, always remain aware that an unanticipated occurrence may frighten your team. A good teamster is able to predict a team's actions, but even the best teamster cannot totally anticipate their reactions. Never put yourself or anyone else in a dangerous position with your team. Mature oxen are so large and powerful that *safety must always come first.* ★

14

COMPETING
WITH
OXEN

Throughout the northeastern United States numerous ox competitions take place, most often in conjunction with agricultural fairs. These events offer ox teamsters a chance to compare their animals, training techniques, and equipment. The draft animals, people, and equipment provide a superb environment for demonstration, questions, and casual experimentation. Competitions have facilitated working oxen remaining in New England in particular because they inspire young people to challenge themselves and adult teamsters to compete in order to challenge their fellow teamsters year after year.

Origin of Competition

Ox competitions have been popular for centuries, historically sponsored by agricultural societies with the primary purpose of sharing information and spreading technology. The original New England ox competitions provided an outlet for fun and entertainment where teamsters could gather and compete with animals normally used for work around the farm or in the forest. Animals were walked many miles to compete in local events.

The contests set standards for local teamsters. Animal training levels, equipment, and techniques all benefited from the challenge of trying to beat your neighbor in friendly competition. Well-matched and beautifully trained teams became the norm, and competitions ultimately raised the level of teamster skills and animal abilities.

Teams were pitted against one another in pulling competitions in order to see who had the strongest team. Classes were put together to see how well-behaved and trained the teams were. Teamsters would hitch the animals to a cart or wagon and maneuver them through a variety of obstacles.

Judges assessed their ability to handle the demands placed on them and the teamster's ability to control them in challenging situations.

Plowing was an important job for oxen on the farm, and many agricultural events included plowing competitions. Classes were later developed to evaluate individual animals and teams based on their appearance, conformation, and how well they were matched.

Early events both praised and scorned the ox teamster. Competitions in Massachusetts that raised the level of teamster skills were praised by agricultural societies, while the same groups often looked down on pulling competitions. In areas where logging was more common, such as the northern forests of New Hampshire and Maine, pulling competitions were more popular. Oxen used in the forest were accustomed to pulling heavy loads and their teamsters were at ease with asking their animals to push themselves to their physical limits. Many early New England documents display the differences in philosophy among — and sometimes outright challenges to — teamsters of other regions.

According to historian Jochen Welsch, who has studied the use of oxen in New England:

OXEN LORE

"Oxen represented the sinews of the region's agricultural strength, and it is no surprise that Yankee farmers took every opportunity to celebrate the animal that had become part of their heritage."

Competitions Today

Ox competitions continue today in the United States, but most animals and teamsters have become specialists. Few teams are used regularly for farm work, logging, competition pulling, obstacle courses, plowing contests, and showing. The teams used for pulling are often well trained and powerful but sometimes lack the desired conformation or other characteristics needed for showing. Show teams are well groomed and well matched but are not usually as well trained or physically conditioned for work and therefore carry extra flesh. Teams used for plowing and log scoot competitions tend to be the least specialized, and both show teams and pulling teams might compete. Only the occasional team used in the field or forest is somewhat specialized in the "old" skills. Even less common is a team that can successfully compete in all four types of event. Some competitions do not allow animals to compete in both shows and pulls, *forcing* teamsters to specialize.

While a good team can do well in all styles of contest, teams trained by 4-H children are the most likely to be able to compete at all levels. This is the result of lots of time and effort spent training, rather than some unique ability or developed skill.

Internationally, competitions have the potential to raise skill levels. Peace Corps volunteers and mission workers tell stories of exceptional teamsters in developing countries who have become the standard toward which local teamsters strive. Such examples are good lessons for development projects or extension workers attempting to introduce oxen and related technologies. Well-trained animals and teamsters will always inspire others to adopt technology and new techniques.

CONFORMATION CLASSES

Most open shows include a conformation class where teams are judged on their conformation, body condition, and how well two animals in a team match. Many teamsters spend a great deal of time selecting animals and feeding them appropriately for this type of show.

Evaluating conformation begins with a careful examination of each animal for serious injuries or

---

Showing and Judging

Oxen are shown in New England as other cattle and livestock are shown in other regions. An ox show normally consists of numerous classes designed to exhibit working cattle that are well trained and well matched as pairs and display outstanding conformation. Competitions also challenge the teams to demonstrate their abilities as work animals.

A single official judges most events, although team judging is sometimes practiced. Both open shows (for teamsters of any age) and 4-H shows (strictly for youth working with their 4-H clubs and projects) are common at agricultural fairs. Shows vary in size and scope depending on the location, interest of ox teamsters, and premiums paid to teamsters for exhibiting. Classes vary with the region and local customs.

Ox teams in New England are expected to work without halters, bridles, nose rings, or other restraints. The cattle are directed with voice commands and the use of a stick or whip.

---

conformational defects that would lower the ox's value in the yoke. The oxen are then evaluated for general appearance. They should have straight toplines, strong straight legs, and strong shoulders and necks. Adequate muscling, condition, and body size for their age and breed are also judged. The hooves should be properly trimmed and free of disease. The animal should be able to move and walk easily.

A challenge in judging one animal against another is that one may have weaknesses or strong points that must be accounted for in comparison with other teams. No set rules or scorecards have been developed for judging the conformation of oxen. A good cattle judge can determine which cattle appear healthiest and most trouble-free, and which display serious conformation problems. Cattle of any breed may be exhibited as oxen, and breed character must be taken into account. Personal preferences about color, breed, or origin should not be part of the judging process.

An ox is not a dairy or beef animal and its conformation should be so judged. It should not be thin and angular like a dairy cow, nor should it be round and fat like a beef steer. A good judge understands that adequate body condition and muscling are essential, but that oxen must display the characteristics of an animal ready for work.

WORKING CLASSES

Working classes are the highlight of any ox show. A well-trained team can display a level of training that inspires awe in the audience. Even farmers who raise beef and dairy cattle may be surprised to see how well oxen respond to subtle cues from their trainer.

A working class has a prescribed course through which the teamster must direct the team. In an obstacle course the team pulls a sled or cart. The best-trained team is one that demonstrates good response to the teamster's cues and that can go through the obstacle course without hitting anything. Yelling, whipping, grabbing the yoke, and pulling or push-

With barely an inch to spare for maneuvering the cart, this 4-Her squeezes her team through.

ing the animals are all frowned on, particularly in a class testing ability to work in the yoke.

Some judges encourage the teamsters to demonstrate any special tricks their animals have been trained to do. The teamster might back the team from behind while not touching the animals, call them to come from a great distance, or take off their yoke and direct them as a team or as individuals. Teamsters have also been known to ride an ox, or direct their team through the obstacle course using voice commands while riding in the sled or cart.

Animals at a show must look their best and be so exhibited as to demonstrate their strengths and downplay their weaknesses, while wearing no halter and using no lead rope.

New England's young teamsters train their steers so well that judges are encouraged to challenge them with seemingly impossible obstacle courses. Nevertheless, many competitors have no trouble meeting such challenges.

BEST-MATCHED CLASSES

Ox teamsters go to great lengths to find two animals that match in color, conformation, horns, weight, and disposition. Many teamsters specifically mate related or even twin cows to a particular bull to maximize their chance of getting two calves that will be similar in all regards when mature. Matching two oxen as closely as possible increases their attractiveness and value, as well as their ability to work more effectively in the yoke. Animals of different breeds, size, or disposition may be a challenge to train to work effectively as a team.

Since beautifully matched teams sometimes lack perfect conformation or working abilities, this class gives teamsters a way to exhibit and compete with animals that may be neither perfect in conformation nor trained to pull a cart. Occasionally, pairs of matched teams are yoked together for four-ox or larger hitches to demonstrate the teamster's ability to select large numbers of matching animals.

FAT OX CLASSES

Oxen were historically worked until they were about eight years old. When the animals began to slow down and had maximized their size and weight, farmers often sent them to market. Colonial New England farmers considered oxen a sound investment because they not only worked for most of their lives but also provided additional income at the end of their working days. Cattle may live beyond the age of eight; many have worked until they were twice that.

The market value of mature cattle in the past was higher than it is today. With this understanding, some agricultural fairs continue to have fat classes to judge the beef characteristics of oxen. Even today, most oxen eventually end up on the table, although the value of fat oxen tends to be ridiculously low in today's market.

Cattle deemed too heavy to place well in conformation or working classes often place high in fat classes. For some ox shows this class provides a mechanism for spreading out the premiums and ribbons among pulling oxen, working oxen, and those considered to be show oxen. Some shows go to the extent of barring any pulling animals from showing, working animals from entering fat classes and fat oxen from competing in other events.

4-H AND YOUTH SHOWS

Youth and 4-H shows allow young people to exhibit their animals in an environment that is less intimidating and often more rewarding than competing against adults. A great advantage to these shows is that each team's performance is the result of the young teamster's own time and effort.

Although most 4-H programs encourage parental leadership, the majority of the work on any project should be done by the youth. Oxen work best for the teamster who trained and understands them. A high purchase price, animals trained by parents, and the selection of a beautifully matched team shown by a youngster who did not put in time and effort readily become obvious in a youth show.

YOUTH SHOW CLASSES

Most youth shows include a number of classes such as fitting and showmanship, best-matched pair, cart or working class, and stoneboat performance class. A conformation class is not usually included.

Oxen are work animals, and training and care are the focus of the 4-H program. Less-than-perfect

---

Seeking a Perfect Match

A best-matched class allows teamsters to exhibit their beautifully matched teams against others. This class tests a teamster's animal selection skills and ability to care for and feed two animals so that they do not grow apart. Most judges evaluate how well the team is matched in color, color patterns, bodyweight, conformation, movement, disposition, horn length, and horn shape.

Judging Youth Shows

The judge's role is to maintain a professional yet challenging environment. Judges should be carefully selected and required to encourage good ox training and showing skills. To assist the youngsters in perfecting their skills, a good judge offers constructive comments and some positive feedback for every young teamster.

conformation should not influence the teamster's ability to compete effectively in working, best-matched, or showmanship classes. A lack of conformation classes allows youngsters to compete on an equal basis even though they may not have chosen their own animals or perhaps could not afford animals with perfect conformation.

4-H shows present every child with a ribbon and prize based on the Danish System, where blue, red, or white ribbons are given depending on the quality of the performance. Blue ribbons are presented

This young teamster has no trouble directing his steers to back a cart while keeping the wheels on 3" planks.

A 4-Her does the seemingly impossible by directing one wheel over a block of wood and stopping on top.

for an excellent performance, red for a good performance, and white for an adequate performance.

A working steer or ox project involves more preparation than other 4-H projects. The animals require not only year-round feeding and care, but also training. That training takes a lot of time, commitment, and planning. Good leadership and lots of encouragement and direction are important for beginning teamsters. Oxen cannot be trained to work a few days before the show. Training must begin months before the show and ideally continue on a daily basis. Young teamsters rise to the occasion by regularly amazing audiences with their skills in training and working oxen.

4-H FITTING AND SHOWMANSHIP CLASSES

The fitting and showmanship class is typically the first class of any 4-H show. Both the animals and the equipment are expected to be immaculately clean and in excellent condition, while the teamster shows the animals to their best advantage. Most teams are clipped to enhance the animals' appearances. Working steers and oxen are supposed to be masculine work animals, so are not clipped like dairy cattle.

The recommended procedure is to body-clip the steers one month before the show, shortening the hair coat yet allowing time for it to grow out and look normal. Animals that are body-clipped just before a show often appear uncomfortable, may be prone to sunburn, and have hair that sticks straight out instead of lying naturally. Many 4-Hers clip their steers or oxen before the show season starts, which lets short new hair grow for the season. This practice eliminates the need for the animals to shed out and makes their coats easier to wash.

Hair in the ears and under the belly and sheath is clipped a few days before the show to give the animal a clean, lean appearance. Some teamsters clip the topline straight to enhance the appearance of the back. Clipping may both enhance strengths and minimize weaknesses.

Regular grooming is important for show animals. Use a bristle brush to remove dirt, dandruff, and loose hair and to accustom the animals to being handled. During warm weather, regularly rinse the

Clipping the coat enhances an animal's appearance by bringing out his strengths and minimizing his weaknesses.

animals with a hose to remove dandruff, dirt, and debris in the haircoat.

Well before the show, completely wash your oxen numerous times with soap to remove manure or tough stains. Just before the show, wash the animals to remove dirt from deep within the coat and enhance the coat's shine. Clean the hooves and horns, and sand and polish the horns.

The grand event for New England 4-Hers is to represent their states at the Eastern States Exposition in Springfield, Massachusetts.

Many farm museums and living history farms sponsor plowing contests to attract the public and offer fun for draft animal owners.

PLOWING CONTESTS

Plowing contests require a judge, a field to be plowed, and rules for the competition. The contest evaluates both the teamster's ability to control the oxen and the team's ability to plow adequately in a timely fashion. Plowing is a difficult job for a team of oxen. Depending on soil conditions, plowing can create a tremendous challenge for the animals to overcome. Plowing contests have for centuries been a way to test both the working ability of cattle and the skills of their teamster.

Plowing was a necessary part of colonial life, so plowing on early farms was an important job for

---

Fitting and Showmanship Scorecard

While no set scorecard or criteria exist, many New England 4-H shows use a scorecard similar to this:

I. Fitting the Team

a. Condition		10 points
b. Uniformity		10 points
c. Cleanliness		20 points
d. Clipping		10 points
e. Hooves		10 points
	Subtotal	60 points

II. Showmanship

a. Appearance of exhibitor		10 points
b. Ability to control and show animals at their best		10 points
c. Teamster's poise, alertness, attitude, and behavior		10 points
d. Knowledge of the project		10 points
	Subtotal	40 points
	TOTAL	100 points

an ox team. The animals had to be willing to pull a plow through rough stony fields, follow a furrow, and work for hours on end. A well-trained team could do the job easily, but a poorly trained team found it difficult. Most oxen had plenty of opportunities to develop their skill and muscles on the job. Today's teams spend much of their time in a pasture or barn, with little or no time spent in the furrow.

Teaching a team to plow requires both good control over the animals and oxen that are ready for the job. Training takes two people, one on the plow and another driving the cattle. As the animals become acclimated to the job they may be trained to plow with just one person to direct the team while controlling the plow. Both plowing with oxen and driving oxen from behind are challenging. The ability to combine the two truly tests the proficiency of both the animals and their trainer.

LOG SCOOT COMPETITIONS

Many agricultural and forestry events feature log scoot competitions. In New England, oxen were once used to twitch logs to a common landing, load them onto a scoot or bobsled, and haul them down a trail. Competition events may combine all three components or consist of just twitching or hauling a scoot through an obstacle course.

If the log scoot competition combines all three parts of a logging operation — twitching, loading and hauling — it is a challenging and lengthy event. These events are judged using time and accuracy in maneuvering the scoot as the basic criteria. Generally the log scoot component is the most difficult because the load is heavy and must be navigated through a tight course consisting of right and left turns and specific starting, stopping, and resting areas. To be fair to younger and smaller teams, most competitions are divided into classes by weight or age.

Ox Pulling

The sport of ox pulling has changed little over the past 150 years. This unique part of New England culture has never drifted far from our stony hills.

Historically, teams were accustomed to long days and heavy loads. They usually spent the winter logging and the warmer months in the fields. Such teams were smaller, leaner, and tougher than the average teams today. They knew what it was like to work 8 to 12 hours a day drawing heavy loads, plowing rocky fields, or walking 20 miles with a heavy cart. Competitions demonstrating the team's skill or strength allowed early farmers and loggers to show off the fruits of their labor.

Plowing Scorecard

Here is a set of possible judging criteria for a plowing contest:

Field Preparation and General Appearance

Bed straight and even, both ends without a marker furrow	10 points
Clean furrows with uniform depth and width	10 points
Furrows firmly against each other	10 points
Land of uniform width when finished	10 points
Minimum disturbance of headland by plow	10 points
Trash well covered	10 point
Subtotal	**60 points**

Teamstering and Sportsmanship

Speed, voice, and driving	10 points
Presentation of self and team	10 points
Team's continuous movement with each furrow cut and good starting and turning on headland	10 points
Staying within own plot	10 points
Subtotal	**40 points**
TOTAL	**100 points**

Cultural differences sometimes arose in competitions, as demonstrated by an article printed in the *New England Farmer* in 1842. A Maine ox man challenged teamsters from Massachusetts with the following statement, after reading that teamsters at the Worcester Cattle Show were hauling two tons of stone on a flat stoneboat.

"Two Tons!! why that isn't a load for a pair of Kennebec calves. We saw Peleg Haines of Readville, at the drawing match at the Kennebec Cattle Show the other day, hitch his single yoke of oxen to a load that weighed six tons five hundred and ninety, and walked them up a hill just as easy as you would a wheelbarrow. When he got to the steepest part of the way, he stopped them a moment, just to show the spectators how easy they could start it again. At the word they moved forward as readily as they did at the bottom — no wringing, twisting, or fuss about it. None of the oxen drew less than 8,500 pounds."

Today ox pulling remains similar to early descriptions. The cast of teamsters, the animals, the yokes, and the keen competition are all the same. The loads the animals pull are tremendous and hard to believe possible. Loads of 15,000 pounds of cement are frequently moved over dry gravel on a flat New England stoneboat and loads in excess of 20,000 pounds over wet gravel. Surely such animals deserve more admiration than they have received.

Most modern-day teamsters consider pulling a sport or hobby. It is a way to work with animals, get together with friends and enjoy a unique cultural competition. Oxen competitions are no longer an integral part of society, nor are teams as common as they once were. Yet in recent years, at fairs such as the Fryeburg Fair in Maine, more than 400 ox teams have been exhibited.

Dozens of competitions occur in Maine, New Hampshire, Vermont, Connecticut, and Massachusetts each year. Competition season begins in early summer and continues until the grand finale at Fryeburg in mid-October. Many teamsters spend the entire season following competitions, some attending up to 50 different events in a season. They take time off from work or use their vacation time to pursue a sport that once involved walking a team to the local fair.

Animals in the contests vary as much as the people who flock to the fairs to watch the pulls. They come in all colors, shapes, and sizes. The largest animals are usually the Chianinas, dwarfing the fair's largest draft horses. At the other extreme is the tiny Dexter, which weighs only 800 to 1,000 pounds at maturity. The old New England favorite has been the Milking Shorthorn, often called Durham by ox teamsters. In recent years Milking Shorthorns have lost ground to more exotic breeds and various crosses.

CLASSES

Pulling teams are divided into classes by weight. The two animals in a team have to meet certain weight requirements, often forcing teamsters to use animals that are not similar in size or color. The weight classes are usually in 400-pound increments beginning with calf classes, where the combined weight of the team does not exceed either 1,200 or 1,600 pounds. Most of these events are limited to children either 14 or 16 years old.

Some fairs prefer not to have calf classes and begin their ox pulls with teams that weigh 2,000 pounds. In years past many pulls used a tape measure to determine weight classes, where the heart-girth of the ox determined the class. Both weight and heartgirth have their disadvantages because competitive teamsters are creative in finding ways to meet the requirements for their preferred weight class or division.

COMPETITIONS

Pulling competitions vary according to local customs, and each has its challenges. Many teamsters prefer just one type of pull and train their cattle accordingly. Other teamsters do well in any style of competition. Exceptional teamsters with exceptional teams are needed to excel in the three main styles of pull.

Distance Pull

In Maine, the distance pull is the most common ox pulling competition. The animals are hitched to a load, ranging from 75 percent to 160 percent of their bodyweight, and are tested on how far they can pull that weight in either 3 or 5 minutes.

Larger cattle, weighing in excess of 3,200 pounds as a team, generally pull the heavier percentages using a 5-minute time limit. These large cattle classes consist of mature oxen with many years of experience. Other big ox classes include the 3,600-pounds class, the 4,000-pounds class, and the free-for-all or sweepstakes teams, which have no upper weight limit. The bigger the ox, the better.

Younger cattle pull a lighter proportion of their weight for a shorter period of time. Many of the "calf" classes are strictly for children, who completely direct and control their animals; other young

Free-for-All

At any fair the free-for-all is the highlight of ox pulling, drawing standing-room only crowds. The gargantuan animals are usually slow and calm, and tower above their teamsters. The loads they draw are tremendous.

people may assist with hitching the team to the stoneboat. Classes start with animals that weigh a combined 1,200 pounds or less as a team. Although these competing teams are just calves, most teamsters will tell you, "the sooner you find out if a steer or young bull has what it takes to pull, the less time you're going to waste on training one that doesn't have the moxie."

Elimination Pull

In New Hampshire, Massachusetts, Connecticut, and Vermont, the 6-foot elimination pull dominates, where each team pulls a loaded stoneboat 6 feet in one continuous motion. Each team is given three attempts. If they fail to pull the load 6 feet in three separate attempts they are eliminated and may not continue to the next load. The winning team draws the heaviest load the farthest distance. In tough competition, none of the teams can move the final load the entire 6 feet. The team that moves it the farthest wins.

The distance pull for adults requires each team to take a load of 100–150 percent of their combined bodyweight and pull it as far as they can in either 3 or 5 minutes.

In the free-for-all class, large oxen, like these Chianinas weighing in excess of 6,000 pounds, pull loads of more than twice their combined weight.

Many of these competitions are held in indoor arenas and tremendous loads are pulled. Free-for-all teams have pulled more than 20,000 pounds of concrete blocks on a stoneboat. With proper training a team can generate huge bursts of energy to pull these enormous loads. To do well the animals in a team have to be in almost perfect sync.

Nova Scotia Pull

A third type of competition sometimes seen in New England is the Nova Scotia pull, where teams from Nova Scotia challenge New England teamsters using Canadian rules. The teams are hitched to one load on a 10-foot-wide track. Each time the team moves the load 3 feet, it qualifies for more weight. The team that pulls the greatest percentage of its own weight 3 feet in a straight line is declared the winner.

TRAINING

Animals used for pulling must have a strong desire and willingness to work in the yoke. A reluctant animal can rarely be made into a champion pulling ox. A good teamster usually has cattle that work easily and pull willingly. If betting came to the ox-pulling ring, the odds would be low because the teamsters' skills play a large role in the ability of the animals.

In the elimination pull, a team has three chances to pull the load 6 feet in one continuous motion.

The Winslow Family
Falmouth, Maine

Not far from the Atlantic Ocean, just north of Portland, Maine, is the Marston homestead, 45 beautiful rolling acres of fields and woods, home of Mark and Kim Winslow and their children Stefan (18), Justin (14), and Marissa (13). Mark grew up across the street from what was then his grandmother's family farm.

Entering their home, it is hard to miss the numerous historic photos hanging on the wall in their home office or ox trophy room, as they more fondly call it. The room has dozens of trophies their children have won with their working steers over the last 10 years. Hanging also are old photos of Mark's grandfather and his Uncle Lester Marston's oxen plowing, haying, and even pulling a car out of the mud in the early 1900s. Mark proudly states that "For six generations oxen have been kept on this farm."

Mark and Kim bought the farm from family and have made many improvements to the house and property. Before moving here, they lived in Raymond, Maine, where they began their family. There Mark rekindled his interest in oxen and got his oldest son Stefan started with a young team. Mark was a 4-Her from 1966 until 1976, a member of the Brass Knobs 4-H club, to which his children now belong. Mark noted that this was just a few years after Bob Young and other parents started one of the first-ever 4-H clubs for working steers and oxen in the early 1960s.

The Winslow children have always impressed me with their teams and the obvious work that has gone into them. Mark and Kim take the kids and steers to about 20 events a year. They travel all over New England to living history farms, plowing contests, farmer's pulls, ox workshops, and, most often, agricultural fairs that host 4-H Working Steer shows.

The kids all started in 4-H when they were 8 years old, and Stefan says he is now retired from 4-H, after 10 successful years. He always had Milking Shorthorns except for his last pair, which were Chianina-Holstein crosses. He will attend the University of Maine majoring in Construction Management, and Mark hopes he'll return to work in the family business.

Justin has had four pairs of steers: two pairs of Milking Shorthorns, one pair of Devons, and his current pair of Devon-Lineback crosses, a striking red color with a narrow white stripe over the rump and tail.

When asked his advice for a new teamster, he replied, "You have to work with them every day, and never give up."

Justin enjoys competing in shows with people from other states and clubs. A serious competitor, he admits he likes to win. He has always demonstrated great sportsmanship in the show ring, however, and when he does not win, he simply goes back home and works his steers to make sure they are ready for the next show.

Mark added, "Four out of five years, Justin has won the plowing match at the Billings Farm, in Woodstock, Vermont," the largest plowing match in New England. (My personal best was second place.)

Marissa has had three pairs of Milking Shorthorns, but admits she would like to have a pair of Devons. Since the family just started a small herd of Devon cattle, I think her wish will come true.

When asked what she likes about the 4-H Working Steer Program Marissa said, "I like showmanship and going to fairs to see who is there. But it is a lot of work."

Mark offers, "The kids get out of these steers what they put into them."

I could not agree more.

As the old adage says, "an ox is not born, he is made." The teamster who knows how to train cattle can get the most out of them. Each animal has his own personality and disposition, and some are easier to make than others. A winning teamster culls oxen that don't have the disposition to pull. Training a steer that enjoys and willingly does his work is easier than forcing an animal to do something he does not enjoy.

Most teamsters start training by walking their animals long distances with a light load. They may then step up the weight and shorten the distance the animals pull. They are building up the team's stamina and wind, then following with muscle and skill building. The greatest mistake many teamsters make is trying to get their animals to pull a large load when they are still young or out of shape. Nothing discourages a young team more than a heavy load they cannot move. The team will remember the experience. Regaining their confidence will take a long time.

Oxen often convince inexperienced teamsters that they cannot possibly pull a certain load. Cattle, like some people, will do the minimal amount of work required. They may refuse to pull out of laziness or they may not know how to draw a heavy load. Many cattle do not realize their own strength until a seasoned teamster provides the right training. Cattle are lazy animals, but once they realize that pulling is part of their life they usually cooperate.

The oldest oxen seen in New England are usually the good pulling cattle. Regular exercise, a controlled diet, and good strong feet and legs help the animals survive to a ripe old age. A common line heard near the pulling ring from old teamsters is "a good team will live long enough to vote." In truth an 18-year-old team is a rarity because the natural life span of cattle is only about 10 to 12 years. Healthy robust cattle of that age are sometimes found, though. Many of the fancy show oxen that are not regularly exercised become overweight, which leads to feet and leg problems and premature culling.

PULLING DEMANDS THE BEST

Training cattle to pull in competition is more challenging than training a team to log or pull a cart. More time, more commitment, and a greater understanding of the beasts are needed. While many young children can train a team to work, few understand how to train a team to compete effectively in New England pulls.

Year after year, the training of pulling cattle inspires teamsters to spend countless hours practicing and getting teams ready for competition. The key to success is a team that is well-conditioned, with lots of stamina, experience in the yoke, and a knowledge of how to push a heavy load upward and forward. A team that learns to "lift" the load reduces the friction on the ground and thus makes the load move a little more easily, giving them the edge in the pulling ring.

A successful competitor like Frank Scruton of Rochester, New Hampshire, with more than 50 years of experience in the pulling ring, says the key to winning is well-trained cattle that put on plenty of miles with a light load. "Once a team knows how to pull heavy loads," says Frank, "you don't have to hitch them heavy very often, just enough so they don't forget. And don't hitch them so heavy at home that you discourage them."

CHAPTER ROUND-UP

Little has been written about working oxen in bygone days, probably because the daily occurrence was taken for granted. The tradition has been handed down through the generations in New England and lives today in the continuation of various competitions. Teamsters who compete in pulling contests have saved and maintained a unique cultural tradition of working oxen. Without these teamsters, oxen and the technology of effectively working them would have been lost in the United States. This valuable living tradition still inspires and challenges youngsters. It is my hope that these young teamsters will continue the tradition into the future. ★

15
KEEPING OXEN HEALTHY

Oxen are easier to keep healthy than are cattle used for breeding and milk or beef production in intensive management and feeding systems. Environment has a great impact on the animals' exposure to disease and conditions that lead to infection, injury, and other problems. Most working cattle are kept in less intensive systems than the methods used for maintaining dairy and beef cattle. As a result, the animals experience less stress, a more forage-based diet, less exposure to disease, and a healthier environment.

Minimizing an animal's exposure to disease organisms goes a long way toward preventing disease. Cattle kept in remote areas or on small farms run less risk of developing disease. Cattle that are frequently moved and exposed to other animals are at a greater risk of contracting and spreading diseases. Young animals have a greater chance than mature cattle of contracting infectious diseases; older animals develop a natural immunity to many diseases.

Normal Signs

Here are some characteristics of healthy cattle:

- Most cattle lie down for eight hours or more each day.

- The noses and eyes of healthy animals are free of mucus or abnormal discharges.

- They have a temperature of about 101.5°F (39°C), with one degree either way considered normal.

- Mature cattle take between 15 and 35 breaths per minute; younger animals have slightly faster respiratory rates.

- The resting heart rate of an adult bovine is about 60 to 75 beats per minute, and calves have a faster pulse.

Unhealthy steers (left) are skinny and losing their hair; by comparison a healthy steer (right) is in good flesh and has a shiny hair coat.

RUMEN CONTRACTIONS

Regular rumen contractions are essential to cattle's well being. Cud-chewing is a good sign that regular contractions are taking place. Cattle should chew their cuds for about eight hours a day.

Other ways to check for rumen contractions are more difficult for the novice. If you think your animal has a problem, check with a veterinarian. With experience, you can learn to watch for movement on the left side, in the hollow spot above the rib cage, and below the hip. Put your hand in this hollow spot and try to feel movement or waves of the rumen wall underneath the skin. Better yet, learn to use a stethoscope to listen for rumen contractions. Ask your veterinarian to help you recognize what is normal and what is not.

Sick or Well?

Awareness of disease symptoms is important when raising cattle. First, understand the normal behavior and appearance of your animals. Healthy cattle have shiny hair coats and attentive eyes and ears. They move easily and rise from lying down with their rear ends first.

Posture varies slightly with the animal's conformation. Regularly observing your individual cattle will aid in determining abnormal posture or movement. Any animal with a definite limp or severely humped back is sick or injured.

Healthy cattle have good flesh covering their bodies, although the amount of flesh varies with the breed. Healthy animals eat aggressively and have yellow urine and semisolid manure. The consistency of the manure varies with diet. Cattle on a high-hay diet have more solid manure. Cattle on lush pasture, silage, or a high grain diet have manure that is somewhat watery, forming a circular pile on the ground or floor.

Cattle are hardy beasts. When an animal becomes severely ill and does not rise, eat, or drink, its condition may deteriorate rapidly. Water consumption and movement are essential to survival. If a sick animal lies on his side for more than 12 hours, you have an emergency situation. The animal is likely to die without prompt medical care.

Administering a Vaccine

Some vaccines for cattle are intramuscular, or given into the muscle. The best place to vaccinate an ox is in the heavy muscles of the rear leg. Do not give it in the neck (as is recommended for beef and dairy cattle), as it could interfere with the yoke.

Most cattle do not take kindly to vaccines. Even well-behaved cattle rarely stand still without being restrained in a chute or tie stall.

I cannot overemphasize the importance of understanding your animals. Many cattle can be infected with numerous organisms that can cause disease without showing outward signs. Healthy cattle have strong immune systems that keep many disease organisms at bay. If the animal is stressed from lack of feed, water, overwork, a new environment, or crowded conditions, he may succumb to disease organisms in his environment. Once a disease strikes, you must recognize the signs early and diagnose the problem before it becomes life-threatening.

Infectious Diseases

Infectious diseases are common to all cattle, particularly those raised near or exposed to other cattle. Most at risk are cattle from sale barns, from large commercial farms, trucked long distances with other cattle, or that attend frequent ox competitions.

Viruses, bacteria, protozoa, fungi, and parasites are the most common causes of infectious disease. They invade the animal and infect some tissue or organ. Some infectious diseases are specific to certain regions; others are rarely seen in healthy cattle. Many of the infectious diseases typically found in lactating cows or cattle used for breeding are not seen in oxen.

The best disease prevention is a clean healthy environment, proper nutrition, and a good vaccination and parasite control program. Most veterinarians are eager to work with cattle owners to develop a program that will prevent diseases specific to the animals and the region where they reside. Despite the many diseases that can attack cattle, they have the ability to develop immunities if they survive initial infection.

VIRAL DISEASES

Viruses are tiny microscopic organisms that invade cells and reprogram them to produce more viruses. The result of this infection is that the cells of the invaded organ or tissue die, thereby creating illness in the animal. Numerous viral diseases infect cattle. Once an animal has contracted a viral disease, treatment is ineffective because viruses rarely respond to medications.

The key to dealing with viral diseases is prevention. Vaccinations boost the immune system to fight particular diseases. Keeping healthy animals separate from infected animals is useful but is sometimes ineffective because insects or wild animals may act as intermediate hosts to transmit a virus. Viral diseases may also be transmitted through contaminated feed or feeders, water, mucus, milk, blood, common needles or other medical equipment, and even manure. Cleanliness and sanitation play a big role in disease prevention.

Following are the more common viral diseases found in oxen.

Rabies. This well-known disease can infect humans. It is caused by a bite from an infected animal and leads to death by paralysis. Vaccinating cattle against rabies offers excellent protection against this disease.

Infectious Bovine Rhinotracheitis (IBR). This respiratory virus frequently infects stressed cattle and is often found when the animals have "shipping fever." A number of organisms causing respiratory problems may be present. IBR may reoccur in animals that have had the disease or have not been vaccinated annually.

Bovine Virus Diarrhea (BVD). Easily transmitted by infected animals through contaminated feed and water, this disease can also pass from an infected cow to her unborn calf. Because the disease affects many body parts, diagnosis can be difficult. Symptoms may include diarrhea and blood in the manure, lameness, coughing, and respiratory infection.

Parainfluenza, Type 3 (PI-3). Although this virus causes only mild disease by itself, it sets the stage for more severe infections from other organisms by weakening the animal's natural defenses, particularly in the respiratory tract. Initial infection may not be noticed until secondary infection occurs. PI-3 is commonly found in calves or cattle that have been stressed by weather, trucking, or poor ventilation.

Bovine Respiratory Syncytial Virus (BRSV). This virus is considered prevalent in cattle populations in the United States. Like other viral diseases, a respiratory tract infection may remain undetected until something stresses the animal and sets off the disease. Cattle may show symptoms as severe as high temperatures, profuse mucous discharge, difficult breathing, or going off feed. This virus may act in conjunction with other viruses or bacteria in the respiratory tract to cause pneumonia.

Rotavirus and Coronavirus. These are two distinct types of virus, either of which may lead to calf scours, the number one killer of calves. Both viruses affect calves at an early age, many at only a week or two. Prevention can begin before the calf is born by vaccinating the cow so that she has antibodies present in her colostrum. A calf must receive colostrum from a cow that has been immunized. In the absence of good colostrum, some commercial mixes provide a modicum of protection or passive immunity to the calf. Once a calf is infected, give him plenty of fluids and an electrolyte solution to help his body fight off the disease.

Warts. Although common on cattle, warts often go unnoticed in commercial herds with light infections. They are usually found only on young cattle and are considered infectious. Even though the animals seem to develop a resistance after they have had warts, many shows are reluctant to accept cattle that have warts. The challenge with warts on working cattle is when they occur in areas like the neck, withers, or head, and interfere with the yoke. If warts become a problem, a veterinarian can remove them. Don't try to cut off a wart yourself, as the wart will continue to grow and cause the infection to become worse.

BACTERIAL DISEASES

Bacteria are larger than viruses and more easily seen under a microscope. These one-celled organisms infect cattle and cause a number of serious diseases. Bacterial diseases are more difficult to vaccinate against than viral diseases, but most bacterial diseases are treatable with antibiotics.

As with viral diseases, animals develop some natural immunity after exposure. Bacteria are common in the environment. A weakened or stressed animal may be constantly re-exposed and develop chronic bacterial infections. Here are the more common bacterial diseases found in oxen.

Tuberculosis. Although largely controlled in the United States by a federal program of eradication, TB has recently resurfaced in both animal and human populations. No treatment exists for this disease in cattle. If animals test positive for tuberculosis, the state or federal government will step in and dispose of the infected animals.

The government intervenes to prevent the disease from spreading from livestock populations to humans. Drinking infected milk or eating uncooked beef was once considered a threat to human health. Because no treatment or vaccine has been found against this disease, the only means of controlling tuberculosis is through testing, retesting, and destroying positively-diagnosed animals.

A recent tuberculosis test for all cattle, including oxen, is important, and the law demands a recent test for cattle taken across state lines. Check with your state veterinarian or department of agriculture for proper testing procedures. Plan in advance of any interstate trip, because the test takes a minimum of 72 hours.

Brucellosis. Primarily of concern in cows, this disease leads to reproductive failure through repeated abortions. I mention it because many states require

testing of any cattle transported between states. Heifers may be vaccinated, but bulls and steers are not vaccinated. Blood tests are conducted on animals that have not been vaccinated. Steers are usually exempt, provided they pass a veterinary inspection and are not from infected herds. Check with your state veterinarian or department of agriculture for requirements and recommendations.

Footrot. A specific organism that invades the hoof and causes severe and rapid lameness causes this disease. Bovine lameness has many causes; footrot is one of the few that may be treated with antibiotics, resulting in rapid recovery.

The disease may appear in cattle with healthy hooves in a relatively dry environment because the organism can infect animals with small lesions or cracks on the hoof that may not be visible. The bacteria typically gain access to the wound or crack at wet areas such as water holes or muddy lanes.

The signs of footrot are swollen pasterns or lower legs, and a great reluctance to walk or put weight on the affected limb. Neglecting these signs can lead to debilitating disease from which the animal may never fully recover. Prompt treatment is necessary for good recovery.

Leptospirosis. Transmitted by wild or domestic animals via urine-contaminated feed or water, this disease is more common in calves than in adult cattle. Fever, jaundice, anemia, and depression are classic signs. Death rates are high in young animals. The disease is difficult to treat. Administration of an annual vaccine is a good preventive.

Listeriosis. Bacteria causing listeriosis are common in the bovine environment and can cause a number of disease problems. The most notorious is an infection of the brain stem leading to "circling disease," in which the animal loses neurological and muscle control of one half of its body.

The disease is common in cattle fed with improperly fermented silage. If the pH of the silage is higher than 5, the organism may be plentiful. Once the disease has progressed to the severe neurological symptoms, full recovery may be impossible. Early diagnosis is important, but often difficult before neurological signs appear.

Lumpy Jaw. Caused by a bacterium that is a normal inhabitant of the bovine mouth and digestive tract, this disease is often seen in young cattle that are shedding their baby teeth. Older animals are easily infected through wounds of the mouth.

Cattle are indiscriminate eaters and may get small wounds in their mouths from consuming coarse roughage, sticks, and other sharp objects. These wounds allow bacteria that normally do not cause problems to grow and cause huge abscess-like swellings of the upper or lower jaw. Early treatment with antibiotics is often successful, but requires accurate diagnosis and good veterinary care.

Navel Infection or Joint Ill. This condition may appear as a simple swelling of the navel, which may go away on its own. Alternatively it may lead to a chronic infection that involves many body parts, most often the joints of young calves. The bacteria that infect a calf shortly after birth may not cause immediate symptoms but remain hidden within the animal's body for months.

The navel can absorb bacteria directly into the bloodstream during or shortly after birth. This disease is easily prevented by dipping the calf's navel with iodine as soon after birth as possible, preferably before he has his first drink. A calf that is born in a clean, dry pasture or stall and has had his navel treated is unlikely to suffer from this disease.

Pinkeye. Although frequently encountered in young cattle on pasture, pinkeye rarely occurs in cattle kept indoors. The bacteria infect the eye, at first causing tears and sensitivity to sunlight. As the disease progresses, ulcers develop and appear as white spots on the pupil. Left untreated, the animals may become permanently blind. Flies may transmit this disease by carrying the bacteria from one animal to another. Tall grass or dusty conditions may further aggravate the condition. Applying antibiotics directly to the eye, if begun early, is usually a successful treatment.

Bacterial Pneumonia. A number of different organisms — *Pasteurella hemolytica*, *P. multocida* and *Haemophilus somnus* — are common bacteria found in cattle with pneumonia. These organisms are normally found in the bovine respiratory tract but cause infection in animals that are stressed, have weakened immune systems from respiratory viruses, or are kept in insufficiently ventilated or crowded conditions. Adequate ventilation, vaccination, and minimizing stress help prevent bacterial pneumonia. Unlike viral organisms leading to respiratory infection, these organisms often respond to antibiotic therapy.

The Dutch Belt steer on the left is healthy and vigorous; the animal on the right is lethargic and not growing well. He suffered from scours as a calf, and soon after this photograph was taken he developed joint ill as a result of not having his navel dipped with iodine at birth.

Scours. Not a disease but a sign of disease, scours is caused by a number of organisms, including viruses, bacteria, coccidia, and other internal parasites. A calf with chronic (long-lasting) and/or severe diarrhea is considered to have this serious, often lethal, condition — the number-one killer of calves before they are weaned. Bacterial scours is common in calves that are bottle-fed. The bacteria that commonly cause scours include *Salmonella* spp and *Escherichia coli*. *E. coli* may be prevented by feeding the newborn calf colostrum from a cow that has developed immunity or been immunized.

Treatment of an infected animal with antibiotics may or may not be effective, depending on the severity of the infection and the level of toxins in the body. Immediate electrolyte and fluid therapy is important for a calf with scours; all too many calves die from fluid loss and low-pH blood long before they succumb to bacterial infection.

Tetanus. Although it is not as commonly seen in cattle as in sheep or horses, tetanus (also known as *lock jaw)* may infect and kill cattle. The bacteria that cause tetanus are found in soil and feces. They are anaerobic and therefore develop in deep puncture wounds. The disease may begin in wounds caused by castration, shoeing, dehorning, umbilical infections, and the application of nose rings or ear tags.

This disease is difficult to treat, partly because diagnosis does not usually occur until it has advanced to a toxic and severe stage. Prevention is as simple as vaccination, avoiding wounds from nails and other sharp objects, and attention to cleanliness before, during, and after surgical procedures.

Parasites

Almost any disease-causing microbe could be considered a parasite because it lives on or in another animal or host. For purposes of this discussion, the word "parasites" will be limited to multicelled animals that infect cattle. Both internal and external parasites can create serious disorders in working animals.

Many new products are available that eliminate most internal and external parasites except for ticks,

Getting colostrum from a cow that has developed immunities or been immunized is important disease protection for a newborn calf.

fungi, and maggots. New products include those that are poured or squirted on the topline of the animal, eliminating the need for oral treatment with bolus guns, oral drenching, or injections. Timing is important with these products, especially for good control of cattle grubs and parasites that are picked up on pasture.

INTERNAL PARASITES

Dozens of internal parasites exist that can cause problems for working cattle. The severity of an infection depends on many factors, including exposure, general health and nutrition, and geographic location. In warm or tropical climates, internal parasites are easily contracted at any time of the year. In cool climates, where pastures are frozen or covered with snow for part of the year, infection rates are usually lower.

Various species of worms and flukes may infect an animal's intestines, lungs, liver, bloodstream, and nervous system. Protozoan diseases like coccidiosis are common in most bovine environments, and calves are much more severely affected than adults.

Symptoms of a calf with heavy infestation of internal parasites include a slow growth rate, unthrifty appearance, poor hair coat, and anemia. The most serious infections lead to death in young animals. Regular internal parasite control is good insurance against severe infections and poor performance.

Animals generally contract such internal parasites as worms and protozoa from contaminated water, feed, and pasture. The typical route of infection is from the feces of one animal to the mouth of another. Minimizing parasites in all animals, and eliminating sources of contamination by maintaining clean water sources and feeding areas, will help prevent infection. All animals benefit from regular deworming and parasite control, for which a great number of veterinary products are available.

Protozoa such as *Giardia* species and the various coccidia species, which cause severe diarrhea and weight loss, do not respond to normal treatment with dewormers. Flukes are common in some areas and are also difficult to control, not normally responding to dewormers. Accurate veterinary diagnosis is important for animals that do not respond to over-the-counter parasite-control products.

The larvae of warble or heel flies cause cattle grubs. The flies lay eggs on the skin of cattle during the summer months. The eggs hatch and the tiny larvae burrow into the skin and appear on the animals' backs in about nine months. The huge grubs cause large swellings and possible infection, and bore breathing holes through the hide. These larvae later escape and drop to the ground where they develop into short-lived but determined flies. The flies do not bite but may severely irritate the cattle on which they lay their eggs.

Cattle infected with internal parasites may appear unhealthy and their performance may suffer. The lungworms and roundworms of the *Ascaris* species that travel through the lungs cause coughing and shortness of breath, severely affect an animal's working performance, and may cause weight loss, anemia, and death.

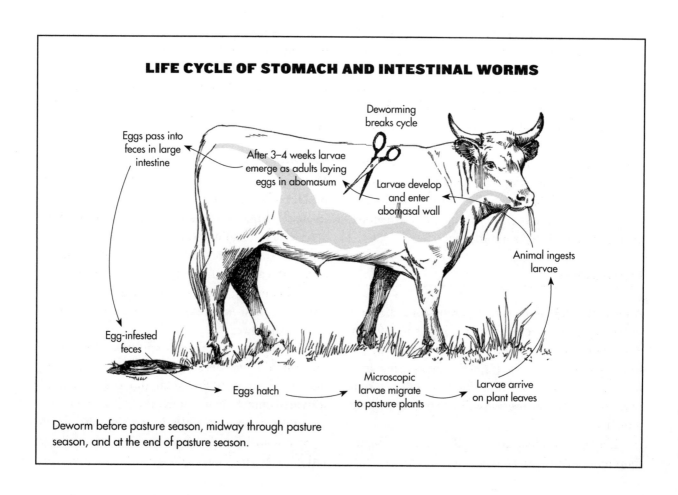

LIFE CYCLE OF STOMACH AND INTESTINAL WORMS

Deworming breaks cycle

Eggs pass into feces in large intestine

After 3–4 weeks larvae emerge as adults laying eggs in abomasum

Larvae develop and enter abomasal wall

Animal ingests larvae

Egg-infested feces

Eggs hatch

Microscopic larvae migrate to pasture plants

Larvae arrive on plant leaves

Deworm before pasture season, midway through pasture season, and at the end of pasture season.

Whatever internal parasites frequent your geographical location, protect your oxen through preventive measures and regular treatment programs. Talk to your veterinarian or collect stool samples and ask your vet to diagnose exactly what parasites are in your herd. Some healthy adult cattle can withstand a high parasite load without adverse effects. The problem is that these high loads act as reservoirs of infection for young or newly acquired cattle.

EXTERNAL PARASITES

Many external parasites affect cattle and may lead to anemia, poor general health, and an animal obsessed with scratching and ridding himself of the parasites. When parasites become the primary concern of working animals, they may fail to respond in the yoke, due to severe discomfort. Treatment for lice, mites, and ringworm is usually successful. Ticks are more difficult to control.

Lice. Although large as external parasites go, lice are barely visible to the human eye. They live on an animal's skin and cause intense itching and hair loss, and in young calves can cause anemia and death. Lice frequently infect cattle housed indoors during cool weather. Animals that are stanchioned or tied and unable to groom themselves often display the greatest discomfort. Cattle in loose housing or wintered in close quarters are also at risk. Treatments include dust-on, injectable, and pour-on products.

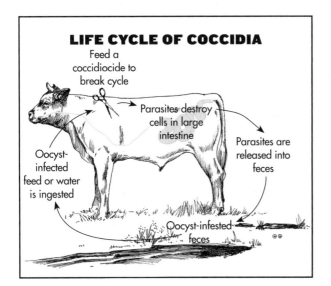

LIFE CYCLE OF COCCIDIA

Feed a coccidiocide to break cycle

Parasites destroy cells in large intestine

Parasites are released into feces

Oocyst-infected feed or water is ingested

Oocyst-infested feces

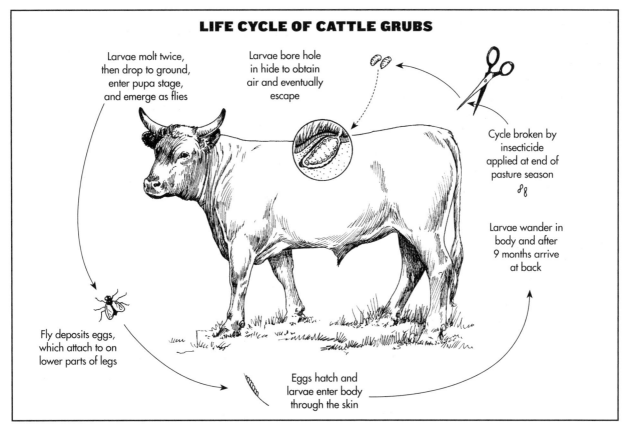

LIFE CYCLE OF CATTLE GRUBS

Larvae molt twice, then drop to ground, enter pupa stage, and emerge as flies

Larvae bore hole in hide to obtain air and eventually escape

Cycle broken by insecticide applied at end of pasture season

Larvae wander in body and after 9 months arrive at back

Fly deposits eggs, which attach to on lower parts of legs

Eggs hatch and larvae enter body through the skin

Mites. Several species of mite infect cattle. They differ from lice in being smaller and in burrowing into the skin rather than living on top of it. Mites cause severe itching, since they burrow through the skin and feed as they go. Most mite infestations occur when cattle are housed during winter, although the parasite is thought to reside on cattle year-around. The most common sites of infestation are the tail-head, tail, thigh, and udder. Signs include intense scratching and licking, often to the point of bleeding. Treatment with commercially available products is usually effective. In a severe case, consult a veterinarian.

Ticks. With numerous species found in many geographic areas, these large parasites attach to their host and gorge on blood, then drop off and reproduce. Severe tick infestations may involve hundreds on an individual animal. Besides being unsightly and causing anemia and discomfort for the animal, many ticks act as intermediate hosts for a variety of disease-causing organisms. The types of tick and the organisms they carry vary from region to region. The bacterial and viral diseases they transmit to cattle include Lyme disease, Q-fever, and anaplasmosis.

Treatment and prevention of ticks includes regular spraying or dipping to kill existing ticks and repel those that attach between treatments. Of all the external parasites, ticks pose one of the greatest challenges. Cattle that are grazing or working in the forest or fields are at greatest risk. Understanding tick life cycles, when they are most common, and ways to avoid them in the field are important preventive measures.

Maggots and Screwworms. These are the larvae of flies of numerous species that lay their eggs on cattle wounds. Once the eggs hatch, the larvae feed on the necrotic (dead) and living flesh. This feeding frenzy may cause wounds to smell of rotting flesh and become full of maggots. The infestation prevents healing and leads to severe infections and seriously ill or dead animals. Prevention begins with regular examination of the animals and the prompt treatment for any open wounds. If maggots appear on a

> ## Dealing with Fungi
>
> Although they cause few common diseases in cattle, fungi are difficult to treat and almost impossible to eradicate. The most common disease affecting working cattle is ringworm, which is easily transmitted to humans. Contrary to its name, ringworm is not caused by a worm but by tiny spores that attach and grow into a fungus that infects hair follicles on the animal's skin.
>
> Ringworm causes intense itching and is easily recognized by round raised crusty areas of skin. Once a barn or stall is contaminated, ringworm is difficult to eradicate. The spores may remain dormant for years to eventually emerge and infect animals housed in close quarters.
>
> Although the disease is most common with animals housed during the winter, it may appear at any time. It normally runs its course in a few months, and infected animals develop a natural immunity for life. Many commercial farms let the disease run its course when the animals are young, ignoring treatment in favor of a good natural immune response. Itching can cause stress for calves and slow growth rates for a short time.
>
> This disease is considered contagious. Animals that are infected are generally not allowed at fairs, shows, pulling contests, or other events because they can be a source of contamination or infection to animals that have not been exposed to ringworm.
>
> Treatment for ringworm is possible. The best medications are available through veterinarians.

wound, several veterinary products are available to prevent further fly strikes and promote healing.

Flies. These can be a nuisance to working cattle, particularly face flies in the field and biting flies (such as deer flies) in the forest. Many effective commercial spray-on preparations are available. For face flies, cattle may be dusted with commercial preparations or provided with a rub where they treat themselves in a paddock or pasture. Ear tags used in commercial herds distract from a working animal's appearance and often rip out; most tags are designed for beef cattle that wear them for only one season. Some teamsters equip each animal with a face net, a simple apparatus the animal can shake to rid itself of face flies while working in the field.

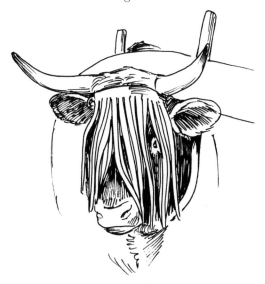

The off ox is wearing a face net, which he can shake to rid his face of flies while working in the field.

Metabolic Diseases

A number of diseases that affect cattle are caused by disturbances in the gut and normal rumen function. Cattle are less susceptible to colic and other diseases that frequently occur in horses; however, rapid changes in diet, improper or inadequate feeding, or going off feed because of infectious disease can lead to serious health problems in cattle. Their large rumens are full of live protozoa, bacteria, and fungi that are essential to health. Any disturbance to this living and essential microbial population will result in at least some compromise to health.

Change your cattle's diet gradually, giving the microbes in the gut time to adjust. In addition, any animal that has been off feed or water, had a high fever, or had huge doses of antibiotics is likely to have setbacks in its normal rumen health.

There are many diseases of metabolic origin. Here are the most common ones seen in working cattle.

Bloat. This condition is a build-up of gas in the rumen. Gas is normally expelled through the animal's mouth in small amounts. When an animal

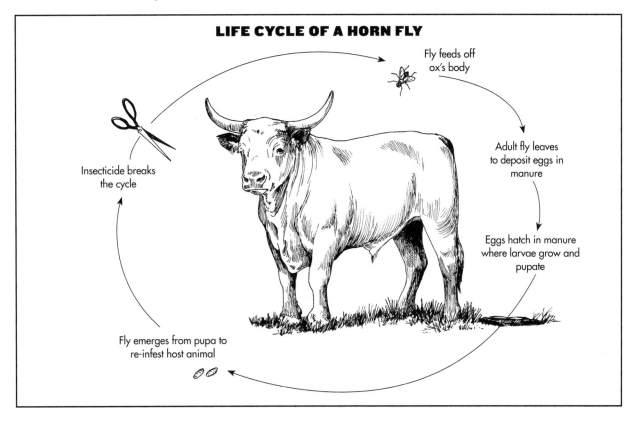

LIFE CYCLE OF A HORN FLY

Fly feeds off ox's body

Adult fly leaves to deposit eggs in manure

Eggs hatch in manure where larvae grow and pupate

Insecticide breaks the cycle

Fly emerges from pupa to re-infest host animal

cannot expel this gas, the pressure builds up to a point that the rumen begins to push against the other organs. This pressure can be so great that the diaphragm, which assists in normal breathing, can no longer function effectively and the animal may suffocate.

Bloat may be caused by rapid changes in the diet, gorging on grain, or grazing on lush pasture (particularly alfalfa and clovers) after spending the winter consuming dry hay or silage. It may also occur in an animal that lies on his side for long periods.

The classic sign of bloat is obvious discomfort, with a large distention of the gut on the left side when the animal is viewed from the rear. As the gases build, the swelling rumen follows the path of least resistance, which is the hollow space forward of the left hip. The rumen may extend to a point where the gut looks like a balloon.

Immediate veterinary assistance is required. The condition is deadly if the animal is not treated promptly. Many drenches are available for treatment, and the pressure may be relieved by tubing the animal.

The old trick of using a trocar and cannula to puncture the animal's side and expel the gas is not usually recommended, especially without veterinary supervision. While it may save the animal from immediate death, poking a hole in the gut and allowing the rumen fluid to enter the abdominal cavity may result in peritonitis (an infection of the body cavity) that kills the animal in a few days.

The best preventions are changing the diet gradually, giving animals on legume pasture a commercial bloat preventive, and carefully monitoring animals on pasture.

Laminitis. Cattle that have become lame without any outward signs of lesions, injury, or disease may have laminitis. Once an animal becomes lame, the disease has already begun to take its course. No quick cure or treatment is known. The sensitive tissue inside the hoof has been disturbed and its attachment to the hoof wall weakened. In mild cases the hoof may repair itself over a course of months.

The normal bovine hoof appears smooth and shiny on the surface, much like that of a young calf.

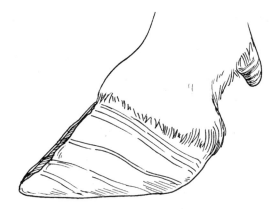

Horizontal lines on the hoof are a sign that the animal has had laminitis.

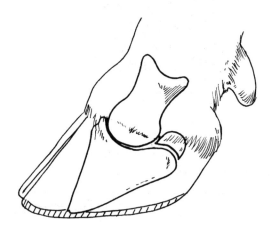

In worst cases the hoof's bones become detached from the hoof wall push through the sole of the foot.

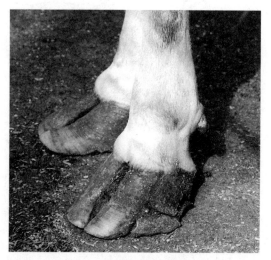

Months after a bout with laminitis, sections of the hoof may break off or crack.

A single sharp line or a few small lines in the hoof tell you that the animal has had laminitis. Deep and visible horizontal lines from hairline to toe are signs the animal has had chronic laminitis. Like a bruise or broken fingernail on a human, these lines will disappear when the hoof grows out to the toe.

Cattle with laminitis may be lame for a few days or, with chronic cases, for months. In the worst cases, when the bones in the hoof have become detached from the hoof wall, the coffin or pedal bone may push through the sole of the foot. Poor-quality hoof horn and a change in the angle of the bone may cause the animal to redirect his bodyweight, resulting in severely distorted claws.

Like founder or laminitis in the horse, the likeliest cause is a rapid change in diet, including lush grass pastures and grain engorgement. Constant heavy grain feeding may result in chronic laminitis. This complex disease is thought to involve inflammation in the hoof and subsequent abnormal blood flow to the hoof, leading to tissue damage. The change in blood flow has been associated with rumen changes caused by going off feed, high fever, or acidic conditions (as when rumen microbes have been disturbed by other metabolic diseases). Working animals are prone to laminitis caused by trauma or injury to the hoof from walking on hard surfaces or wearing their hooves too short.

Laminitis is serious for working cattle. It is easily prevented by providing adequate dietary fiber, not changing the diet too rapidly (particularly when turning animals onto lush pasture), and shoeing oxen when their feet would otherwise wear down too fast. Treatment usually involves anti-inflammatory drugs or painkillers to allow the animal to remain as comfortable as possible during recovery.

Displaced Abomasum. Displaced abomasum is caused by the abomasum rising from the abdominal floor up to a position beside the rumen. In this position the abomasum cannot allow normal feed to flow, and the animal is in serious trouble. The disease is most frequently found in dairy cattle soon after calving. While not as often seen in working cattle as laminitis or bloat, it can be a problem for cattle on restricted diets, those unable to rise because of some other disease, or those on severely limited water.

The classic sign is refusal to eat and failure to defecate, telling you that the gastrointestinal tract is blocked. Veterinary attention is required. Veterinary diagnosis is fairly easy using a stethoscope to listen to the "ping" made by the hollow abomasum against the body wall. Treatment usually involves surgery.

Gastrointestinal Obstructions. Cattle can consume huge quantities of feed in a short period of time. Rapid eating combined with lack of water may cause dry fibrous feeds to impact in numerous locations in the gut. Some cattle are not watered, or may not seek water as often as other livestock. Working cattle are at risk because they may not have regular access to water while working and often consume course dry feed as a large part of their diets. Regular access to water is important, particularly for animals consuming primarily dry feed during the winter.

Obstruction may occur in the esophagus, the abomasum, or the small or large intestine. Obstruction of the lower digestive tract might be characterized as colic because the animal cannot pass feces. An affected animal may go off feed, lie down and get up nervously, or kick at the abdomen. In some mild cases, simply providing a laxative and plenty of water (as a drench if necessary) may remove the blockage. In severe cases, surgery may be an option, but many cattle owners are reluctant to spend money on surgery that cannot guarantee success.

A bovine's indiscriminate swallowing sometimes leads to large objects getting caught in the esophagus. Common examples include apples or sticks, which may have to be removed by a veterinarian.

Hardware Disease. While grazing or feeding, cattle often consume sharp objects that can penetrate the gut and lead to peritonitis. Commonly-consumed objects include nails, screws, and other sharp pieces of metal.

Normal rumen contractions mix and churn the feed in the gut. During this mixing and churning, sharp objects may work their way back to penetrate the front of the reticulum located just behind the diaphragm. Leaking rumen fluid containing normal bacteria and protozoa can quickly cause a life-threatening infection. Even if penetration with leakage does not occur, partial penetration is painful.

An animal with hardware disease may stand with an arched back, move with great difficulty, go off feed, and have a high fever. The only treatment is usually surgery, which may not be cost-effective for working animals. Prevention by administering a cattle magnet at six months of age provides some insurance against sharp objects penetrating the reticulum. A magnet is easily administered with a balling gun, available from any veterinary supply outlet. The magnet remains in the animal's gut for life.

Ulcers. Common in animals on high-grain or silage diets, ulcers may occur in the rumen or the abomasum. As with humans, an ulcer in the gut of an ox makes the animal uncomfortable and, in severe cases, may cause substantial blood loss. The easiest way to prevent and treat ulcers is to make sure your oxen get plenty of long-stemmed forage that requires cud chewing.

Trauma and Injury

Cattle are naturally curious, which can place them in dangerous situations. Working oxen experience more exposure to possible injury and trauma than other cattle. Ill-fitting yokes, poor handling and management, or breakdowns while pulling frequently injure working cattle.

Oxen may be injured in an endless number of ways. Proper training and good animal management in the forest or on the farm will reduce the chance of injury. Make sure your animals are ready for whatever you ask of them.

Yoke-Related Injuries. The neck yoke is a simple but effective piece of equipment if it fits properly and is free of cracks, splinters, and rough areas, especially where it has direct contact with the animals. Yokes and bows that are too large or too small are common causes of injury to the neck and shoulders. An ill-fitting neck yoke or head yoke causes sores, open

Signs of Discomfort

Cattle constantly give clues about the comfort of their yoke. Signs of discomfort include abnormal positions or movement of the head, quickly backing away when trying to pull a load or an unwillingness to work in the yoke. Bows that are adjusted too low cause oxen to throw their heads up in an attempt to find a comfortable seat for the bow. Bows that are too tight cause them to drop their heads to avoid being choked, or swing their heads back and forth to avoid having their necks pinched. Oxen that cough while pulling are wearing improperly fitted bows.

These subtle clues are often ignored until an animal has been injured. It is up to you to make sure your yoke is correctly designed and fitted to your team. Cattle that are properly trained and conditioned for the work pull willingly when comfortable in the yoke.

wounds, or injury. A compromise in comfort results in poor performance and reluctance to work.

Bows that are too low or wide may injure an animal in his shoulders or sternum, or cause permanent nerve damage and muscle atrophy in the shoulders. If the bows are too tight, the yoke will not slide inside the shoulders and will result in "bow bunches," or swellings on the neck.

Without prompt changes in the yoke or an adequate period of healing these swellings may become large, hard, and permanent scar tissue. Such injuries remain with the animal for life. A head yoke may be equally problematic. Poorly fitted and improperly used head yokes may break a horn, crack the skull, or break an animal's neck.

The hitch point affects how the head is carried and how comfortable the ox is at work. A yoke with too high a hitch point pulls heads up. A yoke with too low a hitch point jerks the heads down. Both situations are painful and may injure the animals.

Prevention begins by monitoring the fit and comfort of the yoke every time your animals are worked.

Donald Collins, DVM

Berwick, Maine

"Doc," as my sons call Dr. Don Collins, has been working to keep oxen healthy in New Hampshire, Maine, and Massachusetts for all of his 42 years as a veterinarian. He grew up on a small farm in Berwick, Maine. His dad worked in the woods with draft horses and always had a team of oxen to work as well.

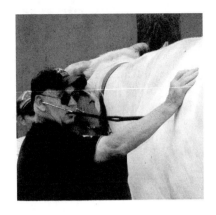

After graduating from Cornell University, Don worked in a mixed veterinary practice with his mentor, Owen Stevens, in nearby South Berwick. He then went out on his own, specializing in large domestic farm animals, and has become well known for his appreciation of and interest in oxen. At least 15 percent of his practice is dedicated to oxen. Ox pulls and agricultural fairs with ox shows are still common in this part of New England, as they have been for more than 200 years.

As haying buddies, Don and I are always talking oxen. He has worked for 17 years on my own animals, doing routine work like castrating, vaccinating, and occasionally diagnosing ailments I could not figure out.

For example, one of my favorite Devons, Tom (star of the film *The Crucible,* in which Daniel Day-Lewis drove him and his mate Buck), developed a swelling on his neck (called a bunch). I had been training and working the team pretty hard all summer. Don took one look at the ox and said, "Give him a rest. That bow bunch is new and soft and there is no permanent damage. If you keep working him, though, he'll have it

for life." I took his advice and the bunch went away. I was impressed with the quick diagnosis, and especially with Don's no-nonsense approach with the team.

The problem Don most often sees? "Sore necks, period!" He explains, "Nine times out of ten the owners have caused the problem, because the steers aren't bowed right. Years ago people used to pay a lot more attention to how the bow fit, but nowadays I bet 40 percent of the owners have no idea how to bow an ox."

When he encounters a sore neck, Don says, "I usually end up giving the teamster a lecture, if he is new to oxen." Having heard many of these, I know his on-farm lectures are helpful, firm, and clear. There are few that can dispute his knowledge and experience with injured or sick oxen.

Don goes on, "Some of the guys who have been pulling a long time slip up and need me to figure out if the damage is permanent or not. In that case, I examine the neck looking for damage to the *Ligamentum Nuchi*, the large ligament in the neck. If the ligament is torn, the ox

is done working. If it is not torn, I can ultrasound it, maybe offer some anti-inflammatory drug, and tell them to give the ox a rest until it heals."

Aside from sore necks, Don says, "The second most common problem in competition pulling oxen is an injured stifle." On his own leg, he shows how the stifle is essentially the same as our kneecap. He indicates how the steer's hock flexes in a heavy pull, and how a hard pull can tear the *anterior cruciate ligament*. This demonstration was a clear example of why farmers understand him and kids are mesmerized by him.

When it comes to oxen, Don has seen it all, and there is no end to his stories. Some days he travels hundreds of miles to work on oxen, and he attends a number of agricultural fairs as the veterinarian on call. His favorite is Fryeburg Fair, where everyone from southern Maine knows him as Dr. Don. He loves to keep abreast of how his clients' oxen do in the pulling ring, and at any fair, if he is not working on an animal he can almost always be found near the pulling ring or ox barn.

Sore necks and muscles are common in steers that are new to working or pulling and have not been gradually conditioned to the task.

Leg Injuries. Oxen that are not trained to pull with a chain may rub their hocks and legs on the chain when turning corners. This rubbing may create open wounds, sores, or permanent swellings on the leg. Extremely sharp turns may result in a broken leg of an unsuspecting ox when one teammate jams the other sideways into the chain. Train your animals to step away from the chain when turning corners, especially when pulling logs or other heavy loads.

Hoof Injuries. The number-one reason oxen in the United States are lost is lameness, most of which originates in the hoof. The most common injuries to the hoof occur when cattle are worked too often on hard surfaces without wearing shoes. Like the horse's hoof, the ox's hoof cannot stand up to abuse on all surfaces. Shoeing adds traction and protects the hoof from excessive wear. Many successful teamsters never shoe their oxen because the amount of work they do and the surfaces they work on do not wear down the hooves faster than they grow.

Another common hoof injury results when a heel catches on the front of a log, sled, stoneboat, or other implement the animal is pulling. You can easily prevent this injury by seeing that your animals have adequate clearance between their hind feet and whatever they are pulling. Young teams generally need more room than older more experienced teams. One steer in a young team will often lag behind the other, and the lagging animal is more likely to catch a foot than the faster or more powerful animal.

Nutritional Diseases

Growing steers and working oxen do not require complex diets, but their diets must be adequately balanced to supply all their protein, energy, fiber, mineral, and vitamin requirements. Although working oxen are easy to feed and care for, they need a balanced diet to meet all their nutritional needs, based on size, age and condition. An inadequate diet not only predisposes an animal to specific problems related to deficiencies but may make the animal more susceptible to infectious and metabolic diseases.

A common problem related to nutrition is poisoning, a problem of particular concern when animals are pastured in an area offering little to eat. Although cattle will avoid many poisonous plants when more desirable plants are available, hungry animals are much more likely to consume plants that may be harmful or deadly. Recognizing the poisonous plants that grow in your area and removing them from your pasture will help prevent this problem.

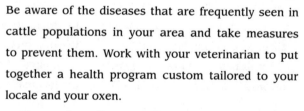

CHAPTER ROUND-UP

Be aware of the diseases that are frequently seen in cattle populations in your area and take measures to prevent them. Work with your veterinarian to put together a health program custom tailored to your locale and your oxen.

Many other ailments affect cattle as well. Some insect-borne infectious diseases found in tropical countries can be deadly. Many parasitic diseases exist and may require special attention or control programs. Individual farms or animals may experience a wide variety of genetic diseases or ailments not mentioned here. The best advice is to consult the numerous texts that have been written about cattle's health problems. ★

16
HOOF CARE

The old adage "no hoof, no horse" applies equally to oxen. Oxen must be able to move comfortably on whatever surface they are required to work. Some oxen require little or no hoof care if they wear their hooves down normally and no faster than the hooves can grow.

Cattle getting little regular exercise tend to have long hooves that require trimming. Oxen that work regularly may wear down their hooves too rapidly and require shoeing. Shoes are needed for pulling competitions and for working on ice, in snow, or on other slippery or extremely hard surfaces such as paved roads.

Restraint

The greatest challenge in trimming bovine hooves is restraining the animal. A foot is easiest to trim or shoe when it is not moving, and in general cattle don't like to have their feet worked on. Their first instinct is to kick. A bovine kick is dangerous. Careful restraint will safely control an animal and allow work on his hooves.

Young cattle may be trained to have their feet held up, but they cannot hold up a foot as easily as horses can. They are (or pretend to be) uncomfortable when holding a foot up for more than a minute or two. A good trimmer can adequately trim a hoof in this time.

An animal that tires of holding up a foot may lie down or lean his weight on the hoof trimmer. Some hoof trimmers are able to hold up a cow or a smaller ox, but a 2,500-pound ox is a different story. A rugged stool or wooden block placed under a front knee will bear the weight of a leaning animal, as an ox can learn to rest his knee on the stool while a front foot is being trimmed. The secret to trimming rear hooves is not to lift the foot too high.

Trimming both front and hind feet without any sort of restraint requires a strong back, excellent trimming skills, and the ability to dodge an occasional kick. Because cattle are more difficult to deal with than horses in this regard, trimming and shoeing oxen have historically been done in a shoeing stock.

SHOEING STOCKS

The shoeing stock consists of a wooden or steel frame with heavy belts designed to support the weight of an ox and allow his feet to be restrained in a position to be worked on. The animal's head is restrained by a halter, by a stanchion, or by tying the horns into the frame of a horn yoke built into the stock.

Once the animal's head is secure, belly bands support his weight. Do not lift the animal up with the belly bands. An ox that is lifted up in shoeing stock will kick and thrash. After an ox becomes used to the procedure, he may relax and slump down on the belly bands.

After the head and the belly bands are secure, lift a foot, by hand or with a rope, and tie it in place. Tie the nearest back foot to the frame so the animal cannot reach up and kick the trimmer. Secure the feet with some type of quick-release knot or other mechanism in case the animal slips out of the sling or the belly bands break.

Once the work is done on the first hoof, untie that foot, let it down, and tie the next one into place on the shoeing stock. Again, tie the closest rear foot to the frame to prevent kicking. Continue the procedure until all four feet are finished.

CHUTES

Many professionals trim dairy cattle hooves using a chute that turns the animal on its side. This type of restraint may be used to trim the hooves of an ox, unless the animal is too big for the chute. A profes-

Stock Size

A **good shoeing stock** is designed to accommodate animals of different sizes, and a great deal of variation exists in the size of working cattle. Animals that are too small for the shoeing stock have too much freedom and may not be safely restrained. Some oxen become so large, on the other hand, that they don't fit into the shoeing stock.

Before designing a stock, explore the possibilities of adjustable footrests, and design your stock with a particular breed in mind. A shoeing stock needed for a Dexter is vastly different in size from a stock needed for a mature Chianina.

Shoeing stock

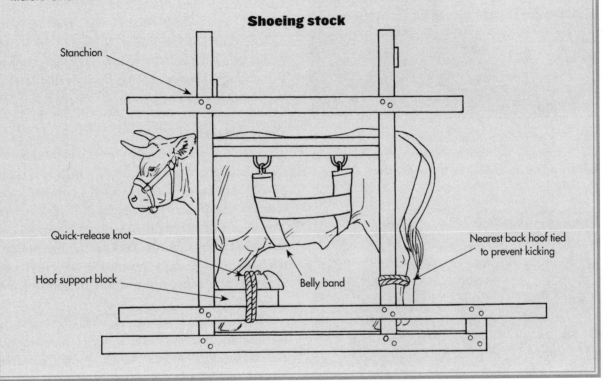

Stanchion

Quick-release knot

Hoof support block

Belly band

Nearest back hoof tied to prevent kicking

---

sional trims the hooves of 20 to 40 animals per day and may not want to come to your farm to trim just one or two oxen.

Hoof Trimming

Regular hoof trimming helps prevent lameness in animals that tend to grow long hooves. Hoof trimming is *difficult, dirty,* and *dangerous.* Cattle not used to having their feet lifted up generally don't like their hooves trimmed, and cattle owners are usually reluctant to trim hooves unless it is absolutely necessary.

Bovines with good conformation and adequate exercise may never need their hooves trimmed. When the hooves grow faster than they are worn down, however, the result is hoof overgrowth. Anyone with more than a few cattle eventually has to trim some feet.

THE HOOF

The normal bovine hoof replaces itself naturally in about one year, which means the hoof grows 4 to 6 inches annually depending on the animal's size. If the hooves are not worn down at about the same rate, the animal must have his feet trimmed. If the

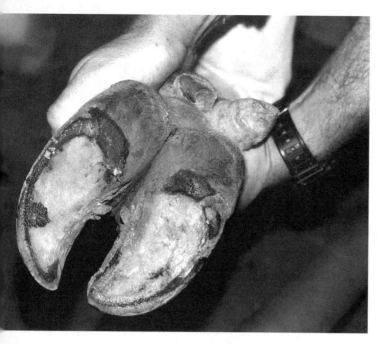

The ideal hoof has two equal claws, a short toe, and a high heel, indicated by the distance between the ground-bearing surface and the hairline just below the dewclaws.

This animal in a trimming chute has distorted front hooves, partly because the front feet are naturally toed out but also because laminitis (as evidenced by the lines on the hooves) caused the animal to walk on his heels.

feet are not trimmed, the toes become too long and force the animal to put weight on his heels, leading to less pressure and wear on the toes, leading in turn to additional and often exaggerated hoof overgrowth.

An ox that walks on his heels is uncomfortable and prone to lameness. He may also develop heel cracks, especially if it is a large ox that is overweight. Heel cracks develop horizontally across the back of the heel and allow foreign materials, including bacteria, to invade the hoof. Bacterial invasion can lead to footrot, abscesses, and other hoof diseases, resulting in a lame ox that possibly ends up being culled. Many working and show oxen in the United States that are otherwise healthy face this problem. Preventing lameness and the diseases that cause it begins with proper hoof care.

The Healthy Hoof

To do a good job trimming hooves you must have a mental picture of the ideal hoof — high in the heel and short in the toe with two equal, or almost equal, claws. A healthy hoof is smooth on the outer surface. The ideal angle of the hoof is about 50 degrees, as measured by putting the hoof on the ground or a flat surface and using a protractor. A healthy hoof protects the sensitive tissue within the hoof and provides traction.

The hoof wall is strongest and thickest near the toe and along the sides of the weight-bearing surface. Most of the animal's weight should be carried on the hoof wall that touches the ground surface. The sole of the hoof protects the sensitive tissue and should not act as a cushion. The sole should bear much less weight than the hoof wall, especially on hard surfaces.

The hoof wall naturally extends beyond the sole toward the ground, with a concave surface, and gives the animal traction. The hoof wall is much like the rim of a cup, and the sole like the cup itself. This cup shape catches debris such as sand, dirt, or snow, adding additional traction when the animal is working.

HOOF AND LEG CONFORMATION

Cattle with desirable hoof and leg conformation are valuable for work because they are naturally predisposed to good hoof health. The ideal hoof does not need regular trimming. When an animal puts most of his weight on the toe, the front of the hoof wears down naturally, keeping the hoof short in the toe and high in the heel.

An animal carrying his weight on his heels will wear down its hooves where they have the least protection. The toe will grow with little or no wear and further exaggerate the problem of overgrown toes. Long toes and short heels cause the animal to exert more energy with every step, and to walk awkwardly and less comfortably.

Leg conformation has a direct impact on hoof shape and wear. An animal that is sickle-hocked will lean back on his heels. An animal that is extremely post-legged will excessively wear down his toes. An animal that is cow-hocked will place more weight on the inside toe and less on the outside, leading to one toe that wears faster and is therefore shorter.

The same holds true for front feet. An animal that toes out generally puts more weight on the inside toe and has a longer toe on the outside. Toeing out could cause larger, more twisted inside toes due to the extra weight on the inside toes.

Cattle with straight legs when viewed from the front or rear, and a desirable angle of the rear leg when viewed from the side, have fewer problems and require less hoof trimming than animals with poor leg conformation.

TRIMMING THE HOOF

Trimming the bovine hoof is not difficult, in theory. The shape of the ideal bovine hoof is your goal, no matter what the hoof looks like when you start. A severely distorted hoof or one that has large pieces torn away cannot be made ideal with one trimming.

The hoof trimmer's job is to restore normal function and shape. If a hoof of a 1,000-pound steer is 12 inches long it may be trimmed back to a normal length of about 5 to 6 inches, provided no diseases or lesions interfere with normal trimming. Visualizing the ideal hoof can be difficult when looking at a severely distorted, overgrown, and dirty hoof. Understanding the normal hoof is essential to knowing how deep to safely trim and rasp.

Practicing on hooves from the local slaughterhouse is a helpful way to develop the skills necessary to trim bovine hooves. Practice will provide you with valuable lessons before you attempt to trim the feet of a live animal. Technique is important in trimming, and speed will come with practice.

Tools of the Trade

The tools for hoof trimming depend on the system of restraint. Electric cutting discs or grinders are not appropriate unless you are a professional hoof trimmer using a tilt chute or shoeing stock. They are definitely not appropriate when trimming hooves to be shod. These tools that can remove a lot of hoof too quickly are dangerous in the hands of a novice.

Every ox teamster should learn to use hoof nippers, a rasp, and a pair of hoof knives (right-hand and left-hand). Purchase the best-quality tools you can afford. Farm supply stores usually have only the cheapest tools available. A good farrier supply catalog will have better designs. The high-quality steel and sharp cutting edges of well-designed nippers and rasps cost more but are worth the money.

Ideally, the hoof should be trimmed from the bottom because it naturally wears on the bottom surface. This procedure makes hoof trimming difficult because the foot has to be picked up for trimming.

Wearing Hooves Down Naturally

Options abound for removing excess hoof growth. Some teamsters find more work for their animals so they can wear their hooves down naturally. This system may work well for animals with hooves that are not too long. Walking on pavement will wear down the hoof faster than on pasture or natural soils, but this often leads to other problems like too much wear, sore feet, or uncomfortable animals. In the neck yoke, oxen may develop the bad habit of pulling away from each other as they try to adjust to walking on an uncomfortable surface.

Some people use a chisel or long-handled hoof nippers to trim the end of the toes while the animal stands up. Clipping the toes, however, does not remove hoof the way it naturally wears down. Proper hoof trimming results in an ideal shape and is more comfortable for the animal.

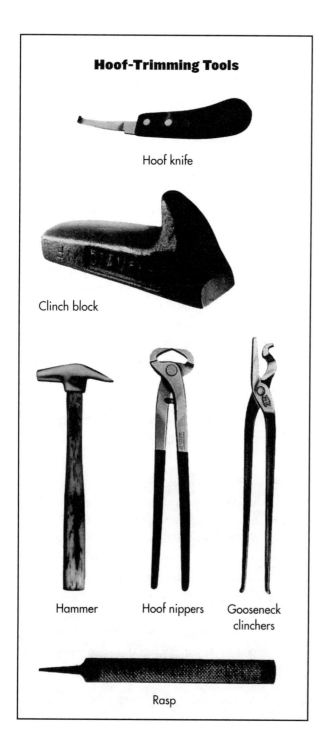

Hoof-Trimming Tools

Hoof knife

Clinch block

Hammer

Hoof nippers

Gooseneck clinchers

Rasp

Restoring the Hoof's Shape

Following are the four steps of removing excess hoof and restoring the ideal shape of the hoof.

1. Cut back the toe. Using the nippers, shorten the toe. The ox may be standing, but trimming (especially the bottom of the hoof) will be easier if the animal is restrained. Some hooves are harder than others. Long-handled nippers make trimming easier if the hooves are hard and dry.

Step 1

2a. Trim away excess hoof wall and sole. Make sure the nippers are perpendicular to the bottom of the hoof. Your goal is to remove hoof as is would naturally wear down. Do not trim the hoof wall from the side or at an angle. The hoof wall is needed to support and protect the hoof. Removing hoof wall and making an ox walk on his soles will cause him to go lame. Correctly trimming the sole is the greatest challenge for the novice hoof trimmer.

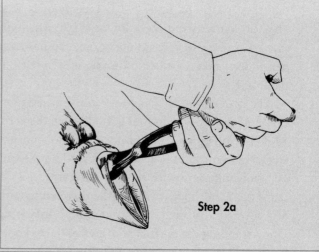

Step 2a

2b. Leave as much of the heel as possible. Remember that the ideal hoof is high in the heel and short in the toe. Rarely does an animal need any heel removed from the back feet and only occasionally from the front feet. Cut deepest at the toe and gradually lift the nippers as you move toward the heel. When using a rasp, work from heel to toe. Put more weight or pressure on the end of the rasp near the toe and remove little or no material from the heel. Proper use of the rasp takes practice.

3. Check the hoof. Release the foot and let the hoof rest on a wood or concrete floor. If the animal is restrained in a shoeing stock, lay a small board or the rasp on the foot's weight-bearing surface. The surface should be level and both claws should be the same shape and size. Level the surface or correct uneven toes, leaving the heel alone as much as possible. Getting the toes even ensures the animal's comfort and natural wearing of the hoof.

4. Slightly cup the hoof sole. After trimming, the hoof wall is usually even with the sole. It should extend slightly beyond the sole. Use the hoof knives to cup the sole so the hoof wall bears most of the weight. Hoof angle shouuld be 50 degrees.

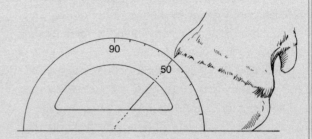

Step 4

The sole need not be shiny and clean. Don't remove more of the sole than is necessary to restore normal hoof function. Thin soles lead to tender feet, especially when an animal is walked on gravel or stony surfaces. Cut out any loose pieces of sole or cracks and pockets that could collect debris and cause lameness.

The only exception to trimming the sole is when the hoof will be shod. In that case, leave as much sole as possible because the shoe requires good contact with both the hoof wall and the sole in order to stay on.

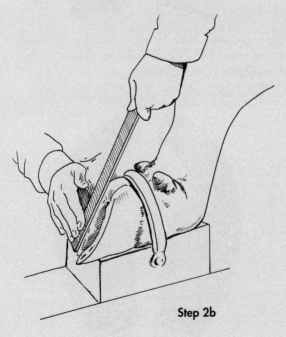

Step 2b

Shoeing

Oxen are shod either to protect their hooves from excessive wear or to give the animals additional traction. Animals that walk on stony or hard surfaces often wear their hooves down faster than the hooves grow. Cattle that are worked on slippery surfaces or used in pulling competitions also benefit from shoeing. Although oxen have been shod for centuries, the difficult challenges of trimming and shoeing cattle may have contributed to their loss in popularity as draft animals.

An ox does not like to stand while being shod. Some teamsters may boast of an ox that will stand on three legs for trimming and shoeing, but these are rare animals. Calm animals will lean on the person who is working on their feet or may lie down to make shoeing impossible. Unruly cattle kick, thrash, and usually get their way. Anyone who regularly shoes oxen finds the investment in a shoeing stock a necessity.

Ox shoeing is difficult to learn without some hands-on experience, such as having previously shod horses. For the beginner, practicing on cattle feet from a slaughterhouse provides excellent training without endangering a live animal. The process of trimming, fitting, and nailing a shoe on a hoof held in a vise is an excellent teaching aid that helps build the skills necessary for shoeing oxen. It also provides a way to develop a sense of how hard the hoof is and how to nail on a shoe. By experimenting and perhaps dissecting the hoof after nailing on a shoe, you will better understand hoof anatomy.

Cattle hooves are different in texture and anatomy from horse hooves. The bovine hoof is more leathery, offering a stronger hold with a lower nail. The hoof wall is thinner, making it easier to quick the animal, or drive a nail into sensitive flesh. The hoof wall is particularly thin in the heel. Cattle that are allowed to wear down their hooves too short present additional shoeing challenges because the sensitive tissue may be close to the inside of the hoof wall or sole. Having two claws on each foot also makes nailing and holding the shoe while nailing difficult.

Shoes cut out of steel, then heated to punch the holes and bend the calks; note the rounder shape for the front claws and the straighter shape for the hind claws.

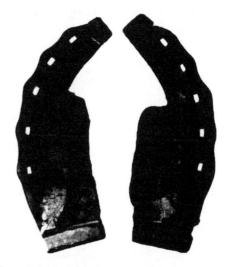

Hand forged shoes made from straight bar stock, heated in a coal forge before the heel area was drawn out and the calks and nail holes punched.

Cast iron shoes are commercially made in three sizes; compared to steel ox shoes they break more easily at the nail holes and tend to wear faster.

To save time and money, some teamsters shoe only the front feet. Shoeing the front feet offers some traction on snow and ice. Oxen used for competition sometimes have the front feet shod because the front hooves wear more quickly in the toe during the frequent workouts necessary to properly condition the animals. Oxen that suffer from excessive wear or are regularly worked on snow and ice should wear shoes on all four feet.

OX SHOES

Oxen wear eight shoes, one on each claw. The type of shoe depends in part on the task at hand. Shoes with calks (sharp-pointed pieces of iron projecting downward from the shoe) are used for additional traction on snow and ice, but create problems for animals walking on pavement and concrete. Flat shoes add traction and protection for animals that frequently walk on hard surfaces, but are little help on snow and ice. In addition, the size and shape of the shoe must correspond to the hoof.

Good ox shoes are hard to find and can be expensive. Ox shoes were traditionally made in a forge by a blacksmith, but today most blacksmiths are unfamiliar with oxen and ox shoes. Modern teamsters find it easier to cut the shoes out of flat steel, weld on calks if needed, and punch or drill the nail holes in a machine shop.

Shoes made of cast iron may be purchased, but they do not wear as well as steel shoes and have the tendency to break more easily. They are also designed for front hooves that are more rounded. Shaping them to fit the more pointed and narrow hind hooves is nearly impossible. Cast-iron shoes are the only commercially available ox shoes in the United States. They may be purchased from Centaur Forge (see the Resources appendix). Shoes are purchased by the pair, four shoes for two feet, or by a set of eight shoes for all four feet.

Shoes are nailed to the outside of the claw. The average ox shoe has more nails per inch than a horseshoe. Most have holes punched for 5 to 7 nails, compared with 10 to 14 nails for a horse's hoof. The nail

Shoe Sizes

Ready-made cast-iron ox shoes come in three sizes. The smallest is the #2, appropriate for animals weighing between 1,000 and 1,200 pounds or larger animals with small hooves. The middle size is the #3, appropriate for animals weighing 1,300 to 1,600 pounds. The largest shoes are #4, appropriate for animals weighing around 2,000 pounds. Mature Chianinas and other large breeds may require shoes larger than #4. These are only guidelines, as the hoof size or shape of each ox is different. A shoe that is too small or too large creates more problems than it prevents.

holes are closer to the outside of the shoe and angled to correspond to the natural shape of the hoof.

Toe nails should be at an angle and heel nails should be more perpendicular to the ground-bearing surface of the shoe. The nail holes are smaller than those used for a similar-sized horse because the ox has a much thinner but more flexible and leathery hoof wall.

Ox shoes are designed to provide more protection and cover on the bottom of the hoof than horseshoes. Although horseshoes have been cut in half and nailed to the feet of oxen, the practice is not appropriate. The ox bears a considerable amount of weight on the bulb of its heel. This weight-bearing surface is important and natural. The heel of the bovine hoof is not like the frog of a horse hoof. It must bear weight with each step.

Most ox shoes are shaped like a comma, for complete coverage at the heel and to correspond to the pointed but curved claw. The coverage on the bottom provides a surface to protect the sole. Some teamsters prefer a shoe that has a slight cup shape under the sole. They believe the cup shape helps hold the shoe on as the sole grows into this concave surface. Most ox shoes, however, are flat and level.

FITTING THE SHOE

Before shoeing an ox, the hooves must be trimmed. Trimming a hoof for shoeing is different from basic trimming. The hoof should be trimmed as level as possible. Do not cup the sole. If the sole has a natural cup, nail the shoe on the hoof wall so that the shoe will be parallel to the ground. Do not nail on a shoe that follows the natural angle or curve of the sole.

The outside of the shoe *must* be made level to fit tight against the bottom of the hoof wall in both the front and back. Likewise the hoof wall must be level to offer a secure and tight fit. The sole will quickly grow and soon fill any void between the shoe and the hoof.

Be sure the shoe fits to the outside of the hoof wall. A shoe that is nailed on with the hoof wall hanging over will create problems for the animal. The hoof wall is the natural weight-bearing surface. Nailing the shoe so it puts pressure on the white line or sole instead of on the hoof wall causes lameness. Even if the animal does not become lame, the hoof wall will grow quickly around and over the shoe.

Make sure the shoe fits to the back of the heel where it starts to curve up toward the leg. If the heel has insufficient coverage, the edge of the shoe creates pressure points that can lead to bruising and lameness. Do not, however, fit the shoe too far back, because the animal is likely to catch the shoe and tear it off. Rear hoofs often catch and tear off front shoes that are fitted too far back. Rear shoes might also catch on things the ox steps on, such as rocks or tree roots.

Fit the shoe to the toe unless the animal's toes tend to overlap. If the shoe is not pointed or small enough in the toe, let the toe overhang about a half-inch, then cut off this extra toe at the end of the shoeing process. This procedure is particularly important for animals shod with calks because sharp calks can

damage the other hoof at the coronary band if the ox interferes (steps on himself) as he moves.

Animals that fight each other in the yoke or pull away from each other often cross-step in trying to maintain their balance. Cross-stepping creates problems when oxen are shod too close to the end of the toe.

If the ox shoe does not fit, do not put it on the animal. A shoe that fits poorly creates problems. Instead, let the ox go barefooted and allow some time off for his feet to grow out. Ox shoeing is challenging enough without creating problems from the start.

NAILING THE SHOE

The hoof wall of an ox is approximately one-third of the thickness of a horse's hoof wall. Oxen are therefore shod using small nails, which reduce the chance of pressure quicking the hoof or splitting it when clinching the nails. Use the smallest nails possible, but get them high enough for a good strong clinch.

City head nails are preferred for small oxen because they are less likely to split the hoof. For smaller oxen use #3 or #4 city head nails. Larger oxen may be shod with nails up to size #6. Compared to large nails, small nails are difficult to nail into the tough hoof wall, but are less likely to damage the hoof. A drop of oil on the nail helps. Do not use rusty nails.

The nails are started at the white line and nailed in much the same manner as for shoeing horses. **Always put the beveled or rough side of the nail head toward the inside of the shoe.** The tip of the nail is also beveled to ensure that it comes out of the hoof.

If the nail is put in backwards with the bevel of the nail head facing out, the bevel of the tip will force the nail into sensitive tissue. An ox shoer has enough to learn without making this mistake.

Put in the middle nails first because they are less likely to pull the shoe toward the back of the hoof, rather than starting with a heel nail, and are easier to put in than toe nails. Nailing toe nails first requires additional support of the foot.

Angle the toe nails more than the heel nails. The hoof wall at the heel is quite thin, so place the heel nail carefully. If the ox has a damaged or torn hoof

Ins and Outs of Nailing

Nailing shoes on oxen offers little room for error. The nail should go in with a fair amount of steady resistance. If a nail goes in easily, remove and reposition it. A tiny bend on the nail's beveled end will help ensure that the nail comes out of the hoof.

Black hooves are usually more difficult to nail than white hooves. This difference influences the feel of the right amount of resistance. Environment and nutrition also create variation in hoof texture and water content. Oxen that are housed or worked in a dry environment generally have harder hooves than those spending their time standing in mud or wet pastures.

Animals on a high-grain diet may have softer hooves or frequently suffer from laminitis compared to cattle consuming primarily long-stemmed fibrous feeds. Knowing your animals' anatomy and getting a feel for nailing is as important as trimming their feet and fitting their shoes.

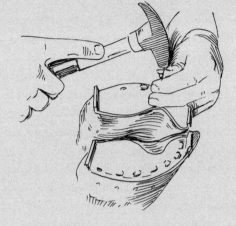

Calked shoes freshly nailed on the claws of a mature Holstein ox. Note the position of nail holes and the extending toes, which will be clipped off when shoeing is complete.

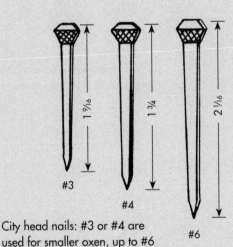

#3 — 1 9/16

#4 — 1 3/4

#6 — 2 1/16

City head nails: #3 or #4 are used for smaller oxen, up to #6 for larger oxen.

Most ox shoers find that the easiest way to properly position a shoe is to put in the middle nails first.

wall, try to get nails where they will hold. Custom shoes work best for an ox that has a damaged or torn hoof wall. Getting a secure fit is essential, but may be impossible if the hoof has been severely damaged. Shoeing an ox with long hooves that may be trimmed to get a good solid wall for shoeing is easier than waiting for the hooves to wear down to a point where the hoof wall will not hold a nail.

CLINCHING

After the shoe is nailed, cut the nails off at the hoof wall and leave about 1/8 to 3/16 inch sticking out of the hoof wall. This excess nail must be clinched to hold the shoe in place. Clinching involves bending the nails over to secure the shoe. Without clinching the shoe will readily come loose and fall off. Every nail needs to be clinched.

Do not place your finger between the hoof and the shoeing stock when the ox has his foot secured. As the ox moves and fights the restraint, your finger could easily be pinched or injured by both the hoof and the protruding nails during clinching.

Before clinching, securely set the nail in the hoof and shoe using a solid steel clinching block. Place the clinching block on the hoof wall above the protruding cut-off nail. Set the nail by giving the head on the bottom of the hoof one or two hard blows. Hold the block above and against the protruding nail on the opposite side to begin the clinch by rolling over the tip of the nail.

Use a hacksaw blade or small file to remove the burr on the underside of the nail, caused when the nail penetrated the hoof. Removing the burr allows the nail to be more securely clinched and less likely to penetrate the hoof wall.

Clinching may be done using a hammer to bend the nail down tight against the hoof wall. Instead of hammering down the clinch, some teamsters use a set of gooseneck clinchers to bend and secure the clinch in one motion. Compared with regular clinchers, gooseneck clinchers are less likely to pull the nail down and tear the hoof, a concern because most oxen are not clinched as high as are horses.

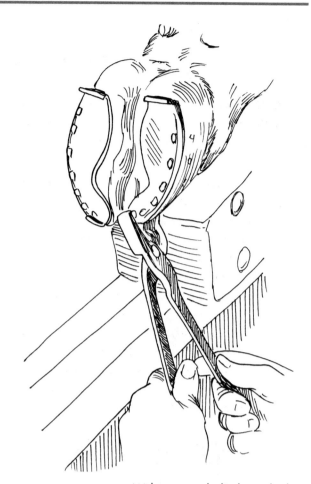

With gooseneck clinchers, clinch the nails to secure the shoe.

After the nails are cut off and set, remove the burr of hoof from under the nail to ensure a strong clinch.

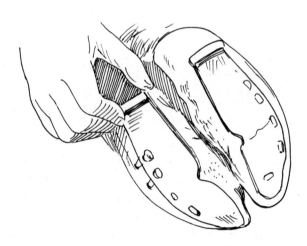

If the nails come loose, or if after six to eight weeks the hoof wall starts to grow around the shoe's edge, it is time to reset the shoe.

Brian Patten
Springfield, New Hampshire

Brian Patten has been a friend since I started with my first team of steers back in the 1970s. Recently, while watching ox pulling with Brian and his wife Kim (Mock) at Ossipee Valley Fair, I realized it had been 30 years since I first met the two of them. Wow, how time flies.

Even if we had not seen each other in all these years, I certainly couldn't forget the Patten family. They have always impressed me with their skills in driving oxen, competing, and being outstanding examples of sportsmanship, combined with a passion for working cattle.

Brian has 40 to 50 cattle, most of them steers, oxen, or cows that he uses to produce bull calves for oxen. Most of the cows are part Chianina, but he admits with a chuckle, "I don't get along so good with the pure white ones (full-blood, all-white Italian-type Chianinas), so I like to breed my own."

Brian's 16-year-old son was sitting with us, and I asked him if he helps his dad much working oxen. Brian is known to work three to four pairs of oxen, all at the same time, to keep his many teams in condition for competition pulling. Tim admits, "I have a hard enough time working a pair or two. Plus, I never did get the hang of working oxen like a train, and my dad gets up earlier than I like to."

Brian is also one of the few people who shoe oxen and will shoe for other teamsters. I asked him to highlight his reasons for ox shoeing and a few pointers on how to do it. He answered, "I try to shoe them only one time per season, and then only put a shoe on as the ox needs it."

What was the toughest thing to learn? Without hesitation, Brian, replied, "Learning to place the nail. Getting the foot prepared for the shoe is probably next. Paring too much away before you shoe is usually the problem."

He had a late start this year and said, "Working the steers like I have, they have more hoof worn off, and it makes shoeing harder."

Why do you shoe? I asked. "Traction mostly," Brian said, looking at the oxen pulling below us. "In a place like this it would have been ideal to have shoes on all the oxen. The other reason is to protect the foot while getting them ready. I do a lot of walking and the steers' feet take a beating [his farm has many rocks and ledge outcroppings]. Some folks do not have to shoe, as they work their cattle on soft ground."

I asked about individual hooves. Brian stated, "Black hooves seem to be tougher. Frank Scruton always said a black and white hoof is about the best, and I have to agree; that seems to be the best foot for shoeing."

How many ox shoers are there? Not many, but we came up with a list of about 15 names. These folks come from New York, Connecticut, Massachusetts, Vermont, New Hampshire, and Maine; however, they do not travel to shoe oxen, and we could not think of any who advertise their services. Some of the ox teamsters who gathered around as we spoke said you'd better not even mention some of these names, because most of them want to shoe only their own oxen.

Brian says he shoes about five teams for other people each year, and he has to charge a price that makes it worthwhile. He has also built some shoeing stocks for other people with oxen and encouraged me to come see his most recent project.

Brian's son Tim admits that if he did learn how to shoe oxen, he would do it for the money. Brian just smiled. He is one of the few people with the skills to shoe oxen, and he knows it is not easy.

Once clinched, the shoe is secure. It should not wiggle against the claw. It should be flat to the hoof and parallel to the ground surface. A set of shoes properly applied should remain secure for about three months. Resetting them may be necessary, depending on the hoof's growth. The shod hoof will grow without wear, creating a smooth flat hoof for the next shoeing.

PROTECTIVE OPTIONS

Numerous products are available that protect the hoof or raise it off the ground. These options are suitable for teamsters who desire an alternative to shoeing for parades, wagon trains, or other activities on hard surfaces. These alternative shoes are designed for dairy cattle with lameness from a hoof disease or other problem. For cattle that are suffering from sore feet and have too little hoof to shoe, gluing on Technovit Blocks or CowSlips may offer a solution. Both products are more easily applied than shoes and usually do not require a shoeing stock.

Technovit is an epoxy that may be placed on the hoof or used to glue a wooden block to the bottom of the hoof. This product offers a temporary solution when ox shoes are out of the question but the animal's hooves need immediate protection from excessive wear. They do not hold up well when oxen are worked.

In most veterinary practices, Technovit has been largely replaced by Cowslips (see the Resources appendix). Designed and manufactured in Ireland, Cowslips are plastic and come in a number of sizes that offer a temporary solution for a teamster with no other options. They are glued onto the sound claw of a pair in order to raise the other injured or diseased claw off the ground. Cowslips may not come in a size appropriate for larger oxen.

Other alternatives include cutting ox shoes out of plastic, rubber, or fiberglass. Such materials offer excellent traction on paved surfaces and cushion an animal from the trauma of walking on hard surfaces. They do not last long on the hoof, however, even when nailed on.

CHAPTER ROUND-UP

Entire textbooks have been written about hoof care and diseases of the bovine hoof. Dairy cattle are the most frequently affected animals, the result of being confined on concrete in wet and unsanitary conditions. Ox teams that are worked in the field or forest and housed in barns, pastures, or corrals are generally free of many of the common diseases of the bovine hoof. ★

17

THE PROBLEM TEAM

Training any animal will have its trying moments, and training oxen is no different. Some teamsters become discouraged or give up when their animals resist the training process. Older animals resist training more than young ones. Working with such animals may challenge even the best teamster. Here are some common training problems and how to deal with them:

Running Away

The most important command is "whoa." Although your oxen should respond to voice alone, they should also respond to visual and physical cues. A team should never learn to overrun the teamster or move without a cue. If your team learns to run away or fails to stop on command it has become a hazard to both people and equipment. This dangerous situation must be remedied immediately.

LIKELY CAUSES

Cattle enjoy their freedom, and most are happy not to work at all. Running away is a sign that the team does not respect the teamster's wishes or system of restraint. Working with oxen is based on dominance and psychological restraint. The team that runs away has learned a great way to upset their teamster. If the teamster gets frustrated and unyokes the animals, they are getting just what they want. The team is training the trainer.

Young steers and disobedient teams should wear halters as
a safety net until they can be trusted to behave in the yoke.

Be consistent in training your team to stop. Do not tolerate animals that don't stop on command. Use exactly the same verbal, visual, and physical cues each time. Expect nothing less than total obedience. Only when you and your team understand the importance of obeying this command should training proceed.

Occasionally oxen are spooked by some unexpected event and may try to bolt. Perhaps the team was not acclimated to people, or the team and teamster were not adequately prepared for an attempted task. Be prepared for the worst at all times. Never take cattle into situations they are not prepared for. A team that has been acclimated by slowly and deliberately being exposed to possibly frightening situations is much less likely to run off in fright.

. . . .

If the teamster gets frustrated and unyokes the animals, they are getting just what they want. The team is training the trainer.

. . . .

Another cause of running away is if the team has previously been beaten and is genuinely afraid of the teamster. When a team is afraid of the teamster the only remedy is to regain the animals' trust without letting them run away again.

CHANGING THE BEHAVIOR

Until the animals can be trusted, keep them on halters and a lead rope even when in the yoke, until you develop confidence in their behavior without the rope. Try to keep the rope slack and use it only as an emergency brake. If the lead is pulled tight to get the animals to stop, then it becomes a cue the animals learn to expect before they will stop.

The team should learn to stop when you do and to watch your motions and whip as well as listen to your commands. Although the lead rope is a good training aid, it should not be necessary if the animals are well trained.

Nose Rings Only a Last Resort

Some teamsters have found the use of nose rings and ropes to be necessary for particularly unruly teams. Nose rings are a severe form of restraint and should be a last resort. Train steers when they are young, work with them often, and be consistent. Follow the rules outlined in chapter 5, and you will not have runaways.

Provide consistent commands and insist that your animals follow them. Oxen may be controlled in many ways, and whatever the system of restraint — ropes, voice, or whip — they must obey the command to stop. Use whatever methods you have to, because nothing is more dangerous than a runaway team.

If you never let your team run away, they will never know they can. If they ever do run away, do not let them get away again. I once had a team trot off when I was riding in the wagon. The team was in a field, and there were not any particular hazards. When the team slowed down, we resumed as if nothing had happened. They never really got away, because I was in the cart the whole time, and they never knew they had momentarily had their freedom.

Ease your team into every task they try for the first time. Preparation and anticipation of what the animals might do is critical to preventing runaways. Many unexpected things can happen while working oxen. Stepping on a beehive, for example, is a dangerous and unpredictable situation. Be ready for whatever happens no matter where you are.

Few oxen that are worked regularly try to run away. If they are tired and hot, running away is usually the last thing on their minds. Work your animals on heavy loads. If the team does not stop easily, work them long and hard so they learn to appreciate stopping and standing still.

Jumping Chain

Oxen must to learn to stay on their respective sides when working on a chain. If one or both oxen get on the wrong side of the pulling chain, they may develop a habit that is hard to break, because they have learned how to avoid working.

LIKELY CAUSES

Animals don't always understand what is desired of them. Before introducing the chain, make sure the oxen are familiar with all the basic commands. Introduce the chain for the first time in such a way that they cannot get tangled up or turned around. Walking them in a straight line is a good way to begin this training. It's also helpful to have someone hold the chain up high enough so the oxen become accustomed to it without stepping on it or over it.

Asking too much of a team of oxen while they are pulling a load may cause them to learn to stop and step over the chain. Even a well-trained team, if asked to pull too heavy a load or pull too far, may jump over the chain in an attempt to get out of the work. If you let this happen, your team may figure out how frustrated they can make you and at the same time get a break from the pull.

Animals with sore necks, pulled muscles, or other ailments may not feel like working. Don't overlook the fact that a well-trained team that jump chain may be telling you that they have a physical problem with the yoke or an injury.

CHANGING THE BEHAVIOR

During initial training, many teamsters use a pole instead of a chain to train the animals to pull. It is helpful to begin training working cattle using a light wagon or sled with a tongue instead of pulling with a chain.

For a young team, the use of a pole creates a training environment where jumping the chain is not an option. If the team jumps the chain frequently even after they are trained to pull, using a pole instead of a chain can help correct the problem.

Another option is to hitch the chain-jumping team between two other teams in tandem so that they cannot get over the chain. The middle team will have the chain in a higher position on their legs, making it nearly impossible to jump over. This ploy

Cattle that learn to turn back or jump chain in trying to get out of work may sometimes be corrected by using a pole, but bad habits are hard to break.

works only when the teams in front and behind are pulling something and are not apt to allow any slack in the chain.

Another training technique is to make jumping the chain uncomfortable by implementing immediate negative reinforcement, but it must be done in a way that tells the animals exactly what they have done wrong. You might, for instance, immediately hit the animal that has jumped over the chain, so he gets back over on his own side. Do not, however, hit a young team just learning how to work with a chain, because they may not understand what they have done wrong.

Still another solution is to force the animals to work with one or both legs on the wrong side of the chain. The chain may scuff up their legs or create further challenges for you as the teamster, but your oxen will learn that it is easier and more comfortable to stay on their respective sides of the chain.

Hauling Out

Oxen are said to be "hauling out" when they pull away from each other while in the yoke. Wooden bows are strong and rigid to lean against. The animals learn that they can pull away from each other by leaning on the outside bow shafts. If one animal leans out the other animal usually counters by leaning out as well. Hauling out is irritating and unsightly because the two animals walk with their legs angled toward each other and their bodies leaning out in an awkward and uncomfortable fashion.

Some oxen pull away from each other so much that it looks as though they will fall over if they lose their footing. The animals may interfere with each other when walking. If they are wearing ox shoes they can tear up each other's pasterns. Besides the possibility of injury to the team and embarrassment to the teamster, the animals are wasting energy that could be better spent working.

LIKELY CAUSES

A yoke that puts the animals too close together sometimes causes hauling out. The animals may be trying to maintain their personal space or avoid their teammate's horns. By angling their bodies and

Hauling out, or pulling away from each other in the yoke, wastes a lot of energy, can cause injury to the animals, and is embarrassing for the teamster.

heads out, long horned cattle avoid being poked by their teammate.

Some oxen initiate this behavior when one animal tries to maintain good footing on slippery pavement or ice. Once initiated the behavior can quickly become a habit. If one ox leans out, the other animal counters this by leaning out too. Frequent walking on pavement and hard surfaces is associated with this problem. Some teams resume normal walking as soon as they have gained good footing. Others are delighted to have found a new way to irritate each other and their teamster.

Certain breeds, like the Brown Swiss, Chianina, and American Milking Devon, more commonly pull away from each other in the yoke than do other breeds. These are intelligent breeds noted for mischievous behavior.

CHANGING THE BEHAVIOR

Make sure the yoke is long enough to allow your team to move and work together comfortably. A yoke that puts the animals three times the bow width apart is sufficient unless they have exceptionally long horns.

Changing the yoke to one that allows more space between the animals is a possible solution. Animals in Africa using the withers yoke don't seem to develop the problem of hauling out because the withers yoke is so loose and flexible that the animals cannot lean against it to pull away from each other. Using a head yoke will eliminate the problem, but is a radical change for the teamster and the animals if they are accustomed to a neck yoke.

Another possible solution is to switch the sides on which the teammates are yoked. Switching sides in the yoke causes slight confusion and the animals don't know which way to lean. They may also be temporarily confused about how to turn and behave in the yoke. Once they become accustomed to working on either side, switching them back and forth seems to prevent hauling out, but this may be easier said than done.

Pulling the heads of the oxen closer together while the team is in the yoke forces them to stand up straight. Wrap a short, light chain around the top of each ox's head just below the horns. Pull their heads together so the heads turn slightly toward each other. Chaining the heads together will irritate your animals and force them to walk upright.

Turning the Yoke

Turning the yoke is not a common problem but occasionally manifests itself when animals are just beginning their training. The teammates learn to swing their rear ends away from each other so the yoke turns over. The team may learn the habit to avoid work, since they must be unyoked to correct the problem.

LIKELY CAUSES

Turning the yoke is one of the less dangerous bad habits that oxen develop. Trained teams that are left alone or allowed to stand idle for long periods learn creative ways to irritate their trainer enough to let them loose. If you are not careful about keeping your animals' rear ends together they may discover turning the yoke while learning to back up. Oxen learning to back up may swing their rear ends so far away from each other that they inadvertently turn the yoke.

Be careful to keep your animals' rear ends together or they may discover turning the yoke while learning to back up.

CHANGING THE BEHAVIOR

Avoid this problem by never leaving your team idle and unattended in the yoke for long periods. Make sure the bows are tight enough so that the animals don't have the flexibility to twist and rotate the yoke comfortably.

Another solution is to tie the animal's tail switches together with string or the long hair on the tail. When the animals pull away from each other, it pinches their tails and creates discomfort, which can stop the behavior. Watch your animals closely and correct them immediately if they do turn their yoke.

Lying Down in the Yoke

Occasionally one or both animals will attempt to lie down and refuse to get up and work. Young calves that are tired may lie down, and this can be expected. Older animals may learn to lie down to resist training. Be careful that your animals do not make it a habit.

LIKELY CAUSES

Calves less than four months old tire easily. At agricultural shows held during warm months in New England young calves often try to lie down in the yoke. More than resisting the training process, they are usually just tired. Rest your animals between classes or allow them plenty of time to rest before a show. They usually grow out of it as they become older.

This problem is troublesome and occurs frequently in developing countries where oxen are regularly pushed to their physical or psychological limits. Training that does not allow adequate acclimation or conditioning is often the cause. Animals are sometimes expected to plow after 11 months of idleness or only a few days of initial training.

An ox team trained to plow must be conditioned to this heavy work. Oxen faced with severe punishments and an unclear idea of what is expected of them may simply give up. In most cases it is not a physical collapse because of exhaustion, but a psychological collapse because of confusion. The teamster is pushing the animals beyond their limits of knowledge.

Another possible cause is initial training that began with severe restraints. Examples include the excessive use of nose rings, use of a running W to stop wild animals, or dragging animals with a tractors or another ox team. Even the wildest team will have their spirit broken by such severe restraints. The result, however, may be far less desirable than the result of bringing an animal along more slowly with less severe systems of training and conditioning.

In developing countries oxen are often beaten or rushed in their training, resulting in animals that are reluctant to work.

A calf may lie down in the yoke just because he is tired after a long day in the show ring.

CHANGING THE BEHAVIOR

The simplest way to keep an animal from lying down in the yoke is never to beat your animals or push them beyond their physical abilities. Training and physical conditioning are important. If an animal is hot, tired, sore, and constantly pushed to do more than he feels he can, lying down is his last defense.

If your animals have been worked to near exhaustion, unyoke them, rest them, and try the task again. Since they may learn to expect unyoking and resting when they lie down, take care to prevent them from lying down when they are not tired. If your animals try to lie down, prevent this by keeping them moving. Oxen that find a way to get out of working will remember this self-reinforcing trick for a long time.

Make sure their equipment fits properly. A poorly fitted yoke that causes discomfort and sores may be the reason why trained animals resist work. Check your equipment and your animals to guard against a sore, injury, or poorly fitted yoke.

Two animals that differ in strength, size, or temperament may be the problem. Yoking such animals with new mates to see if they work better is a possible solution. Oxen have unique temperaments. When all else fails, changing teammates is worth a try.

Kicking

Cattle can kick swiftly and in almost any direction with incredible accuracy. Kicking is dangerous behavior you should never tolerate or encourage.

LIKELY CAUSES

What may have begun as a fearful kick can soon become a way to attain dominance over humans. Cattle use their horns in defense and their feet out of fear. Most cattle first run from what they fear, then use their heads if necessary. Kicking is a last resort used most often in close quarters or when the head is tied.

An ox that learns that he can keep people at bay with his feet will do so. An initial fear of trainers can become a self-reinforcing behavior that keeps the animal from being yoked or handled, which may be just what he wants.

Any harsh training that encourages an animal to kick is a poor training technique. The more the animal kicks and keeps people at bay, the more difficult curbing this behavior will be. Sometimes an ox will learn to take advantage of certain people and not others. It is no different from the pecking order in a herd. Cattle get away with whatever they can.

CHANGING THE BEHAVIOR

Kicking is a difficult behavior to curb if the animal has been successful at keeping people at bay. A kicking ox often gets what he wants. If you can identify the first aggressive kick and instantly administer negative reinforcement, the animal will readily see that kicking you or other humans doesn't make sense.

If an ox seems to have taken up kicking for no apparent reason, house and work the animal in a way that makes kicking impossible. Avoid the behavior or stimulus that initiates the kicking. Harsh and quick methods rarely lead to animals that are trustworthy and calm. If you are rushing their training, ease up on your animals. Instead try to brush and calm them in close quarters so they learn to trust their handler.

Aggressiveness

Like kicking, aggressiveness is dangerous behavior that cannot be tolerated. A calf's playful head butting turns into a dangerous game as the animal matures. Never play around with cattle and encourage head butting or charging. Never tolerate the slightest push with the head or horns.

LIKELY CAUSES

Cattle fight with their horns. Testing their strength and power by pushing each other is natural. They frequently mock-fight, and such mock fights often turn into real battles for a better position in the herd's pecking order. Similarly, cattle may try to test their strength on their trainers or other people. This testing with the head should never be encouraged or tolerated.

If just once an animal realizes his strength and potential power when he uses his horns, he will learn to use his aggressiveness to establish dominance over humans. Never tolerate even a bump from a horn. A subordinate animal would never challenge the boss cow or bull with even the slightest poke of a horn. The same rule must be applied between the ox teamster and the team.

An ox can learn to establish his dominance over certain people, especially children, by bluffing them with his head and horns. Sometimes this is done without the teamster knowing about it. When animals are on pasture or tied in the barn, they may learn that they can scare some people and not others. This selective aggression is normal in the bovine world. When the ox sees people jump away or run, he has learned he can dominate some humans.

CHANGING THE BEHAVIOR

Do not tolerate aggressive behavior. Just as it is natural for the ox to test his power, it is also natural for him to be subordinate to more powerful or intelligent animals. Humans can control oxen because the animals perceive us as dominant. In the pasture or corral an ox respects only strength and aggressive behavior from his bovine counterparts. He will work only if he believes you are tougher or more powerful.

Establish dominance while your animals are young, rather than waiting until they have established some status or enjoyed their freedom in the field for a few years. More power, intelligence, and stronger ropes and fences would be needed to establish dominance over a 2,000-pound ox that has never been handled compared to an 80-pound calf that is fed in a stall.

Once an ox has established himself as dominant over humans, the only way to reestablish control is to show him who is boss. One way is to use a larger ox to force the animal to work in the yoke.

Another way to reestablish dominance is through a carefully planned but extremely dangerous training session, where the ox will lose his superior status. Such a session involves severe restraints or throwing the animal. Predicting just how the animal will react is difficult. Some give in, some lose their spirit,

and others fight harder. An older ox that has learned all the tricks of being dominant over humans may not yield easily to just anyone, especially someone smaller and less aggressive than those he might respect. If one person reestablishes dominance in such a session, the ox may respect only that one person, as he would respect the one cow that is the boss in a herd.

Humans can control oxen because the animals perceive us as dominant.

Avoid this problem by never allowing your oxen to take the upper hand from the beginning of their training. If your oxen will be exhibited at fairs, contests, parades, and museums, get rid of any aggressive animals. Bringing an ox back to total submission is difficult once he has realized his power. Cattle can recognize individual people. An aggressive ox that is made to respect humans through force may temporarily give in but may never be trustworthy with the person he once took advantage of.

Inattentiveness

Animals that have been trained and handled by many people often learn to ignore commands from individuals they feel are not likely to force them to do something. The oxen may respond only when they have to or ignore commands altogether. This gesture of independence is normal but should not be tolerated.

LIKELY CAUSES

Animals learn best when reinforcement follows a certain behavior or cue. The fastest learning takes place when the animals receive consistent visual, verbal, and physical cues from their trainer. Too many commands given at one time, too many distractions, or a trainer who is unfamiliar with the normal routines associated with working oxen can result in uncooperative animals. An animal that is sick or exhausted will also fail to respond to commands.

Wild cattle or animals that have had little time in the yoke may have no idea of what the teamster wants. Early in training they will not understand what it is the teamster desires. Their inattentiveness will be remedied only by time spent being trained and worked in the yoke.

CHANGING THE BEHAVIOR

Animals that do not behave as expected, but have just begun the training process, often come around to learn what is expected of them. It takes time to train oxen. Teamsters who follow the basic rules, regardless of their level of experience, have few problems. Provide an environment that challenges your animals, but do not push them beyond their level of training.

> Teamsters who follow the basic rules, regardless of their level of experience, have few problems.

Mixing up your training routine to include a variety of real tasks and destinations will help make you and your team more familiar with each other. Remember that you are in command at all times. Your team has no choice but to follow your directions. Training is safer and easier if you never break this golden rule: You make the right choices and you give good directions.

Reluctance to Pull

Cattle can develop sore muscles just as human athletes do when they push themselves beyond their level of conditioning. The neck and shoulder muscles are usually the first places oxen become visibly sore. Sometimes during a heavy pull, an ox will attempt to push into the load, then quickly snap back as he pulls or strains a muscle.

LIKELY CAUSES

Oxen need to build their muscles and become acclimated to the work they will be doing. Rushing into heavy work without prior physical conditioning may result in sore necks and animals that are reluctant to work. As your oxen begin their training, condition them slowly to every physical task required of them.

Poorly fitted yokes and bows can lead to breakdowns when animals are asked to do difficult tasks like logging, plowing, or competition pulling. Poorly designed yokes or bows that do not fit will lead to sore shoulders and necks.

An animal with a sore neck will be reluctant to pull no matter how persuasive the teamster might be. This reluctance is most often seen in animals that are normally willing to work in the yoke, but after a heavy pull appear completely unwilling to work. A sore neck is uncommon. Do not confuse an animal's balkiness caused by poor training with an animal with a sore neck. Many oxen try to get out of work if they think they can.

CHANGING THE BEHAVIOR

Rebuilding the animal's confidence is different from changing rebellious behavior. Allow him adequate time to rest and recuperate. If the muscles are sore from hard work, a few days may be needed before he is ready to work again. For a severe injury, the recuperation time may be weeks, or even months. When training resumes, gradually build up your animals to every difficult task required of them.

Not Working as a Team

In every team one animal is stronger or faster than the other. Every teamster must face the task of getting two animals to work as one, but the animals must learn. They will eventually come together if they are directed to do so and are worked regularly.

LIKELY CAUSES

No two animals are exactly alike: one ox in the pair is always dominant. The dominant animal may want to walk ahead of the other. It is up to the teamster to teach the oxen to walk in unison by slowing one animal or speeding up the other. Once your animals

In most teams, one ox is stronger than the other. Unless they learn to work together, the stronger ox may push the team in a circle or knock down the weaker, slower ox in a heavy pull.

begin working the dominant ox may or may not be the most aggressive and harder-working of the two. The ox teamster's job is to get the animals to start together and maintain the same pace.

CHANGING THE BEHAVIOR

Making two animals walk together at your pace begins the first day halter training begins. The sooner your animals learn to walk at the desired pace the more readily they will learn to walk and work together. Coax one animal to speed up by calling his name and tapping him on the rump. Slow one animal down by tapping him on the head. Most oxen easily learn what is desired of them.

Even with regular coaxing an animal sometimes continues to work at his own pace. Careful selection of animals in a team and later selection of their positions in the yoke have some impact on how animals work together. When animals are of different breeds or temperaments, one may never be comfortable working at a pace set by the other. As a result, one animal will always hang back and let the other do most of the work.

Put the larger, stronger, or faster ox on the off side, so that he pushes the slower ox closer to you, where you can constantly coax him along. If you put the strong ox on the nigh side, he may push the slower animal away from you and the team may wander away.

When two animals learn to come together during a difficult pull, but continue to walk at different paces in the yoke, you may have to learn to live with it. When the opposite is true, accomplishing work with the team is more difficult. Two animals that will not walk together during a pull cannot walk a straight line when plowing. Worse, one animal may be knocked down or off balance by the faster or stronger animal. If you cannot remedy this problem, finding a new teammate may be easier than constantly badgering a slow or extremely fast animal.

Grazing in the Yoke

Cattle working in the field or forest are attracted to everything lush and green they see, especially if they are on a limited diet, as is often the case with mature animals. Oxen attempting to graze in the yoke are exercising their natural desire to eat.

LIKELY CAUSES

Grazing in the yoke distracts animals from their work and eventually leads to a failure to pay attention to the teamster's desires or cues. The animals may become obsessed with eating everything in their path instead of following the furrow or row of crops.

While they have their heads down to graze, the oxen are at risk of being overrun by a heavy wagon or getting the yoke ring or chain caught on their horns. If the animals are being used for weeding or cultivation, they may eat the crop as they go, reducing the harvest with every step. The simple solution is never to allow your animals to graze while in the yoke.

CHANGING THE BEHAVIOR

Keeping your animals attentive is most easily accomplished by using muzzles or nose baskets. After a few failed attempts to graze, muzzled oxen usually give in and pay attention to their teamster. Muzzles are especially helpful when oxen are working between rows of a particularly tempting crop, like corn.

Muzzles should be designed to allow the animals to breathe easily and even drink when allowed to do so. Muzzles made of fine wire mesh, poultry fencing, or welded wire work best. Rope or leather straps fashioned into ox muzzles can work but tend to limit the animals' ability to breathe easily and dissipate body heat.

Oxen may be allowed to graze or feed before a workout to diminish their desire to eat once in the yoke. This ploy usually works with animals that are worked for only a short time each day. When an animal reaches to graze, tap him on the nose or head and give the command "head up." Before long the well-fed team will learn to keep their heads up.

CHAPTER ROUND-UP

Training oxen is full of challenges. Ox teamsters must get the best out of what they have to work with. When training a team, maintain control, strive toward perfection in every regard, and never let your team get the best of you. ★

Kevin Daly
Lyndonville, Vermont

I first met Kevin a few years ago when I visited his family's dairy farm in Lyndonville, Vermont. A high-school student, Kevin had a barn full of steers and oxen he was training for pulling that summer. Serious about competing, he was getting ready to take his teams to a fair that evening.

Now 20, Kevin is a student in the Forest Technology program at the University of New Hampshire. He has had steers and oxen for 13 to 14 years. His dad, Mike Daly, had pulling oxen years ago. Kevin says, "I got started by getting dragged around by a pair of Herefords in New York, when we lived there."

While at a fair watching an ox pull, I asked Kevin about solving common problems teamsters face in training their teams. He said, "They will always come along if you are patient. Some steers just get whatever you are training them to do, the first time. Others might take five or ten tries to get the same result."

Asked about specific training problems, Kevin said, "If they are jumping chain, get them around as quick as you can." With hauling out, "Sometimes if I get after one of my red ones, he will lean out. I have always been told to get off the road if they haul, and if they get hauling bad, you can switch sides with them [in the yoke]." As we watched, a team entered the ring. One steer's head was tied with a small chain that pulled it in toward the yoke. Kevin commented, "That works, too."

He has four pairs now, three pairs he is pulling and one pair of Chianina-Charolais calves he is just starting. "I got 'em when they were calves. If a new person was in the stall, they would come after them and kick and try to horn them. They just seemed to blow up with anyone new. So I took it easy, and I have been bringing them along slow." He said they are coming along but are still pretty shy around other people.

Every teamster faces challenges in training a team. Getting past the problem, Kevin said, "just depends on being patient and how much time you are willing to put into them. With young steers that do not know much, I give them more time than I will an older ox that has developed bad habits. I had a nigh bull last year and at the first fair he wouldn't hitch. So I worked with him for a week or two, to make sure he hooked. At the next fair, he was back to not wanting to hitch. So he was done, his luck had run out."

Tied to Kevin's trailer at the fair were his 2,400-pound team of Shorthorns, Rock and Bull. They had just had a nice pull but were beaten by some teams that had a quicker start. Kevin pointed out his favorite of the other four oxen tied to the trailer, a big 10-year-old black steer named X, in the 2,800-pound class with his mate.

"He didn't even have a name," Kevin said. "He was the extra bull calf we had one spring when we had about a half-dozen Devon-Holstein-cross calves on our dairy farm. One day I just grabbed him by mistake, as a calf, threw him in a yoke, and he turned out to be the best one I have ever had. So the name X, short for Extra, stuck."

An experienced ox teamster can tell which ox has potential for pulling early on. But as Kevin says you never know, with competition oxen, which ones will work out. Some start out well and get sour; others develop problems; and then a steer like X comes along, and you would never guess that he would be the only one of the six calves that worked out after all these years.

18

OXEN

IN

HISTORY

History changed with the beginning of farming and the domestication of animals. Following centuries of human labor, the first draft animal put to work in early agriculture was the ox, which was used long before horses and other equines were domesticated. Oxen provided the draft power that helped create an agricultural revolution by allowing farmers to till more land, harvest crops in a more timely manner, and transport crops and other goods in large quantities across great distances.

The use of oxen coincided with the invention of the wheel and wheeled transport. Oxen paved the way for the development of new crafts and trades, as well as larger human communities. The importance of oxen in providing power for agricultural development is often forgotten in nations that have long since adopted modern forms of agricultural mechanization.

Oxen in Early History

The first cattle were most likely domesticated in southeastern Europe and western Asia, probably in Greece and Turkey. Cattle husbandry in that region was common 7,000 years ago. Other than domestication of the dog, domestication of cattle was the most important step in manipulating the animal world and exploiting land for agricultural purposes. Cattle supplied the meat, milk, leather, manure, and power for early agriculture.

Cattle were herded to grazing areas and corralled at night to prevent predator attacks. This early system of management ensured frequent contact with humans, making the animals easier to manage and train. It affected the cattle by limiting their feed consumption, thus limiting their growth and size. Early domestic cattle were small compared to today's breeds. The first step in genetic selection was castration. Breeds came about as various cultures and regions emphasized certain coat colors, horn shape and size, and other physical characteristics.

Offering a living history lesson, the Maasai in East Africa today continue to herd their cattle the same as many early Asian and Eastern European cattle owners did. They keep animals that are smaller than American, Asian, or European breeds, much like the nondescript early cattle kept by other civilizations.

The Maasai and others like them in Africa are beginning to adopt a more sedentary form of agriculture, and oxen are an important part of that change. They are just beginning to face the trials of training and using oxen with few purchased inputs, no wheeled vehicles, and often a plow as the farm's sole piece of agricultural equipment.

With the help of their oxen, the Maasai have begun their own agricultural revolution. Planting crops, reducing the number of cattle kept, and

OXEN LORE

In *Tall Trees, Tough Men,* Robert Pike beautifully stated the importance of oxen in the early New England forests and the reasons they preceded other draft animals in almost every agricultural culture in the world.

"Before the horse there was the ox. The ox has many advantages: he is stronger than a horse; he is less apt to be scared; he is less inclined to flounder in snow and mud; he is not so given to sickness; he is less expensive to buy and keep; and if it becomes necessary he is better to eat."

This Maasai farmer is just adopting oxen and uses them as they have been used for thousands of years in other parts of the world.

A boy hauls water to town with his team as a way to both help his mother and help his family generate income.

A Maasai farmer uses his teams to clear land of thorn bushes with which he will build a corral.

limiting travel to distant grazing areas are just a few of the technological changes that oxen have brought to human civilization for centuries. In addition to the ox's most important role in timely land preparation at the beginning of the plowing season, the animal has been adopted as a means of transport. Using crude sleds, the Maasai haul manure to fields, carry water from distant water sources, and move materials from the forest to build fences and new homes.

To these farmers and many like them in developing nations, the loss of an ox or a team may mean life or death. Failure to get crops planted during the narrow window of the rainy season may lead to famine. The loss of work animals increases the need for human labor for daily chores like hauling water, firewood, and other agricultural commodities and inputs. Observing these people and their use of oxen is like traveling back in time. The use, importance, and ability of oxen to bring about more labor-efficient crop farming has changed little in 2,000 years.

The ox is frequently mentioned in the Bible and other early writings, attesting to the animal's importance and working ability. According to early Roman scholar Columella in the first century AD, the ox was a person's most hardworking associate in agriculture. The ancients respected oxen so much that it was capital crime to kill an ox.

Columella said that oxen selected for work should be "young, square, red or brindle, with wide foreheads, broad chests, and huge shoulders. Breaking to the yoke should take place between the ages of three and five." Columella preferred the neck yoke, but recognized the head yoke as the most common yoke in most Roman provinces. The use of head yokes continues in much of southern, eastern, and central Europe today.

Almost 2,000 years later, it may no longer be a crime to kill an ox, but the criteria Columella spelled out for selecting and yoking oxen hold true in many cultures today.

According to Robert Trow-Smith, writing in *A History of British Livestock Husbandry to 1700*:

"Ignorant of the cultivation of hay as a crop, the Medieval farmer was caught in a three-pronged dilemma. To plant his crop he needed draught [draft] power. To make his grain grow he needed manure. Yet, to get his manure he had to have animals which lived part of the year on the meagre crops he raised. It was a vicious unending struggle for oxen, manure and food. The ox was normally working in the fields at age four and kept in the yoke for another six years."

Oxen in the Middle Ages

During Europe's Middle Ages oxen continued to be valued possessions. The early Europeans considered the ox at his prime until the age of 10. The animal was then sold for beef while still in good condition, at maximum size and weight, and provided a nice return for his owner.

Not until the invention of the horse collar and harness, along with the development of large breeds of horse, did the shadow begin to draw over the ox and lead to his decline in numbers and importance. The step from ox power to horse power is seen by many historians as important in the development of civilization. The large and improved breeds of horse were both faster and more powerful than cattle, allowing more rapid transportation and larger individual farms and plots. The change from ox power to horse power was not universal, nor even a phenomenon that can easily be explained.

John Langdon wrote an entire text, called *Horses, Oxen and Technological Innovation*, about the transition between ox and horse power in England between the years 1066 and 1500. According to Langdon, this time period saw a definite trend of the increased use of horses. Some regions did not follow the trend because of the varying wealth of farmers; the type of soil, with wet, clay, lowland or stony soils being predominately ox areas; the availability of pasture; the distance from large markets; and cultural biases. The use of oxen continued well into the nineteenth century, even in England where the bias against their use had existed for 800 years.

Oxen in and from Europe

The end of the Middle Ages brought about the expansion of European culture into the Americas, and later into Australia, Asia, and Africa. Early explorers brought to many parts of the world their cattle, their favorite yoking systems, and a new form of agriculture. From Europe came two distinct systems of yoking and training oxen: the neck yoke and the head yoke, arising respectively out of England and Spain.

The English brought their "middle horned" (mid-sized horn) cattle and neck yokes to Virginia and the New England colonies, later taking them to Australia, and finally to East Africa. The Dutch brought neck yokes to New York and later to South Africa. The Spanish brought their longhorn cattle and head yokes to North America. The Portuguese and later the Germans influenced South America by introducing the head yoke system in Brazil, one of the few nations to use both head yokes and neck yokes, the latter brought with Zebu cattle from India.

The use of oxen and yoking systems spread with cultures. The influence of oxen and their yokes is like the languages and other unique influences that followed the various cultures into new lands and remain to this day. Everywhere people settled in the New World, oxen followed. The lands were tilled to produce crops and harvest trees and other natural resources for the growing populations.

In many areas of Europe, oxen continued to be worked into the early twentieth century. England and the nearby nations of northwestern Europe, which used various forms of the neck yoke, were among the first nations to replace oxen almost exclusively with horses by the middle of the nineteenth century. France, Italy, Portugal, Spain, and much of middle Europe, which used primarily head yokes, replaced oxen with horses by the beginning of the twentieth century. Only a few nations in southern Europe use any oxen today. In Eastern Europe, the fall of the Soviet empire has left many small farmers with few options, resulting in renewed interest in the use of oxen.

Oxen in North America

At the end of the Middle Ages, while European farmers were having great debates on whether or not to use oxen, early American pioneers had no other choice but to use oxen. Manufactured resources were scarce. Harnesses, fancy carts, and good roads would not be available for decades. Early colonists knew cattle could survive under conditions that would kill a horse. Cattle provided the power, leather, milk, meat, and manure to build our nation. Many cattle in northern settlements died because of severe shortages of winter feed. Cattle ate anything they were given or allowed to find that resembled roughage.

Oxen were common in early Spanish settlements in areas that are now part of the United States. In Arizona, New Mexico, California, and Florida, ox teams were a familiar sight by the middle 1500s. Due to Spanish influence, these oxen were used in

A pair of Hereford crosses hauling firewood to the railway station in North Weare, New Hampshire, in the early 1900s.

head yokes. The Spanish Longhorns were all legs, horns and life. By the 1800s, huge longhorn herds had grown in Mexico and the Spanish province of Texas. The herds were so large that some of their owners lost track of their animals. These herds were later developed into the Texas Longhorn, a hardy breed that is well adapted to the harsh conditions of the hot southern plains.

The cattle that survived those early years had few needs but provided the power for tremendous tasks. Teams provided the much-needed power to tame a new land and begin agricultural endeavors. Oxen were used for logging, moving stones, pulling stumps, plowing and harrowing fields, making roads, and hauling carts or sleds filled with anything the early colonist purchased, sold, or moved. Oxen pulled covered bridges into place, moved large buildings, and hauled everything from wood products to processed goods. As lands were developed west of the colonies, farmers packed their belongings into wagons drawn by oxen and began the great Westward Expansion.

. . . .

Teams provided the much-needed power to tame a new land and begin agricultural endeavors.

. . . .

The importance of oxen on farms continued throughout the eighteenth and much of the nineteenth centuries. The ox's prominence may be seen in many photographs depicting early settlement, agriculture, and transportation. Horses later replaced oxen in many northern and western areas; mules became popular in many southern states. Oxen, however, were usually the first to clear the way for civilization. The plodding, patient beasts forded rivers and crossed the hot plains, deserts, and swamps. Oxen provided the power to clear virgin forests, plow virgin soils, and build roads and bridges necessary to bring people to the newly settled areas.

On the Oregon Trail and earlier treks, cattle were preferred over horses or mules because they were cheaper and less likely to be stolen by the Native

Americans. Oxen were easier to capture on horseback and to control in deep mud and water. If they died or were injured, they were better to eat. Oxen could graze at night and gorge on local forages in just a few hours, later to chew cud and thus process roughage better than could horses.

Even though oxen were slower than well-fed equines, on difficult trails they could equal the performance of horses or mules by staying in better condition and setting a slow but steady pace. On good days, oxen could average 16 to 18 miles by traveling in the cool of the morning, resting during the midday sun, and resuming the trek in the afternoon. Depending on the trail, an ox can average 2 to 3 miles per hour and always performs better in cool weather.

As trails were improved, farmland was settled and tough virgin sods were opened in the Midwest, as well as forts and trading posts being established in the West, equines gained the advantage and oxen started to lose prominence in much of the West. Horses or mules became more plentiful and quickly replaced oxen, first in hot regions like the Southwest and later throughout much of the Midwest and West. In the great forests of the Northwest, oxen began what horses and machines eventually finished.

OXEN IN NEW ENGLAND

Preferring cool weather and having no fear of water, oxen were worked in the wet soils and swamps of New England. New England farmers continued to maintain the ox long after much of America, with the exception of a few hill farmers in the highlands of North Carolina and Georgia, had given up on him. The ox was well-suited to the needs of New England farmers. His slow pace and patient manner made him less likely to break farm implements pulled through rocky hillside fields.

The ox was fed little and worked hard. He thrived where other beasts failed and became the farm animal that symbolizes the New England colonies. The tradition of working cattle was passed along from generation to generation. Throughout colonial history, oxen toiled under the yoke, ready for any task demanded of them. By the late seventeenth century, oxen were in every New England community where soil had to be tilled and materials moved on land. Their work was much like the work done by the truck or tractor today. Oxen were not the fastest mode of transportation, but they were readily available, dependable, and cost-effective.

The New England teamster has always been well known for his oxen and his ability to work and train a team. This pair of Devons (photographed with their owner in Freedom, New Hampshire, in the 1930s) is wearing a New England slide yoke, brass horn balls, and shoes on all four feet.

New England's ox teamsters became famous for their work with draft cattle. Englishman William Strickland, passing through Springfield, Massachusetts, in 1794, commented,

"Stout able bodied oxen are everywhere, used for the purposes of husbandry. Hardly a horse is kept upon a farm: two or four go on a plough, a waine or waggon; they all work with common yokes, and are more tractable and better broke than I ever saw in England; it is not uncommon in this country for a boy to begin working a pair in a very light carriage, when they are little more than calves, which it may be intended for him to drive many years afterwards, by which means they become so habituated to him that they will follow him about like a dog.

The Yankee ox-driver is no less character than any of these, and his oxen not less remarkable in their performance, it is believed than the horse. They often pet and almost caress them while calves, yoke them when yearlings, at two years old they will work on a lead, at three can manage a cart, at four, will canter into the barn with a load of hay. From then on, a yoke of them is a pretty good team for almost any purpose. . . .

Have you ever seen fifty or a hundred yokes of these oxen with their drivers, drawing a large building? How they will go down to their work! Scratching, pulling, down on their knees, and up, and down again, knocking their horns, and tearing the very earth beneath them, as if they would go through their yokes."

A couple of East Weare, New Hampshire, farmers in about 1909 take a break from plowing to rest their Devons (in front, wearing muzzles) and Ayrshires or Shorthorns (behind).

These oxen in slide yokes are being driven down the road by a farmer riding his sled in North Weare, New Hampshire, about 1896. Those are Devons in front and blue roans (likely Holstein-Shorthorn crosses) in the rear. By the look of their brisk pace, they are likely on their way home.

Three Hereford teams rolling snow at Weare Center, New Hampshire, in the early 1900s. Oxen continued to be used for snow-rolling long after horses became popular, perhaps because they are not bothered by deep snow.

Oxen symbolize the New England colonies, as evidenced by the author and his team on the 1995 movie set for *The Crucible*.

Museums and living history farms use oxen for special events as an important lesson for those who have forgotten what farming in this country was like less than a century ago.

Although oxen were once important draft animals throughout the United States, New England is the only region that boasts numerous oxen today. After almost 400 years, the ox is still found at work in New England's fields and forests. His presence may seem unimportant, but it is unique. New England has hundreds of ox teamsters whose skills may be attributed to their stubborn forefathers who refused to give up the oxen that had served them so well.

Today oxen are most often seen at country fairs and in parades or historic events in rural communities. Teamsters range in age from young children driving calves in 4-H events to adults competing in log skidding, plowing, and pulling contests with animals exceeding 3,000 pounds each. People from around the nation and the world come to these events to learn skills that have long been lost in other regions.

Oxen may be found in every New England state. Each year their numbers hold steady or grow slightly. The animals remain slow, strong, and dependable.

The training techniques and commands used to control them have remained unchanged for centuries. Teams still wear the simple wooden neck yoke and are still driven with a small stick or whip. Ox teams will no doubt be here for future generations to come.

CHAPTER ROUND-UP

The ox remains an important work animal. At the turn of this century, some 250 to 300 million cattle were employed around the world, representing four times more cattle than equines used for work. The ox continues to grow in numbers and importance in sub-Saharan Africa but has lost much of his importance in more developed regions of the world. In the United States, Canada, Western Europe, and Australia, oxen are kept mainly by small farmers and hobbyists. ★

Darin Tschopp
Williamsburg, Virginia

Darin has been in the field of history, specifically agricultural history, for 18 years. As a history major at Marycrest College, Darin took an internship at the Living History Farms, in Urbandale, Iowa. This began his career.

That career included moving east to Massachusetts, to become the Farm Artisan at Plimoth Plantation, then moving south to Virginia, where he served as an Interpreter at George Washington's Mount Vernon. Later he became an Agricultural Specialist at the Jamestowne Settlement in Virginia.

In his current position Darin is the "oxman" at the Colonial Williamsburg Foundation (CWF) in Virginia, driving three pairs of large Milking Shorthorn oxen and one single Randall Lineback ox on an almost daily basis. Darin started working at Colonial Williamsburg as a Livestock Husbander in 2000. Two years later, after a retirement in the Coach and Livestock Department, he took the position of oxman, and his work with the teams officially began in May 2002.

As part of his professional development in handling oxen for the CWF, Darin has been a participant in four of the annual autumn Ox Workshops that Tim Huppe and I have coordinated in New Hampshire. A reserved man, Darin always watched what we did, and how we did it, with much more intensity than first-time ox drovers. I was impressed the first day when Darin drove our largest teams with ease. However, given his height, being 6 foot 7 inches, I quickly realized he was not comfortable with my short Devon calves. He later admitted Devons were not his favorite for other reasons.

I wondered after the first workshop if Darin would ever come back, because he was so quiet and reserved. I remember Tim Huppe assuring me he would, as he had a great discussion with Darin after the workshop at BerryBrook Ox Supply.

I recently asked Darin why he enjoyed coming all the way to New Hampshire for an ox workshop, and he had a long list of reasons, including:

"The individual attention from all the instructors, and the tips on how to get a team to do something they don't want to do. Watching how other people handle a team when they have a problem. The guest speakers have been great. The evenings when we [all workshop participants] are hanging out and telling stories. The diversity of the participants in the workshops. The great food! [And finally, . . .] the workshops have given me the chance to talk to and learn from experienced drovers that I can't get here in Virginia."

Reflecting on his career Darin said, "I have covered over 400 years of agriculture in our country, along with knowing British agriculture in the late 1500s, early 1600s, because of Plimoth and Jamestowne, and 1700s with my current position. The nice thing about being the CWF's oxman is that my research can focus on oxen. I don't limit myself to the eighteenth century. I look at all time periods. . . . I look anywhere and everywhere for oxen information."

Not being quiet or reserved myself, I cannot wait to have him come back to our next ox workshop, where I will grill him for some of this historic information.

19

INTERNATIONAL DEVELOPMENT

Oxen have been an important part of human development for centuries. They still greatly outnumber horses and donkeys as beasts of burden, and the discussion of oxen as a backward way of farming continues today as it has for centuries. Horses are faster and tractors are more efficient where fuel is cheap and spare parts are available. But for the poor farmer who is far from able to afford either of these technologies, the ox continues to pull the plow. Oxen plug along tilling more soil than any other beast on earth. Their work is often accomplished in crude and uncomfortable yokes with poor training and accompanied by whipping and cussing.

Oxen are just one cog in the wheel of developing agricultural systems. Their use in agricultural development has many faces. It also has many different degrees of adoption, use, and interest. No single simple solution to agricultural development exists for the entire world.

Developing Countries

It may seem farfetched that in some parts of the world, oxen are just now being adopted and put to work. In sub-Saharan Africa they have been used for only about 100 years. In Africa, the agricultural use of oxen is on the upswing. Many farmers there have never seen oxen at work; others remain skeptical about their use.

In Latin America, the use of oxen continues as it has in the past, often side by side with tractors and modern equipment. The great disparity between the rich and the poor means that farmers might work all week on a large farm driving tractor, only to return to their villages and use oxen or donkeys at home. In Cuba, oxen were readopted by farmers, in part by a government mandate, when the former Soviet Union collapsed in the late 1980s, and the farmers had to rely on locally produced agricultural power sources.

Oxen are used for transport and farm work in the rural areas of many developing countries.

Similarly, in Eastern Europe after the collapse of the Soviet Union, many farmers have reverted to animal power out of necessity. These farmers once had tractors and huge farms supported by the central government. Once they had to farm on small plots of their own, the use of oxen was an economic necessity rather than a step toward modern development.

RELEVANT TECHNOLOGY

The topic of oxen and animal power in international agricultural development is closely related to the use and history of oxen in the United States. The relevance may seem far removed from the needs of a farmer on the other side of the world who is struggling to feed his family. True, certain aspects may not be relevant to development workers and animal traction experts, but many of the methods for working oxen have remained unchanged for hundreds of years and can be used to generate new ideas or to perfect old ones.

The techniques for training and using oxen are similar everywhere. Most oxen are worked in pairs. When extra power is needed the pairs are hitched in tandem. The single ox is used in many areas, but not as commonly as oxen worked in pairs. Most oxen are used with a yoke. Although yoke designs differ, the comfort of the animal and maximizing his power are always a concern. The way oxen are trained to work in the yoke and pull a plow or cart is similar everywhere. All oxen must learn to stop and go on command, pull carts or farm implements, turn right and left, and back up, or at least position themselves to be hitched to a load.

When I have seen oxen at work in other nations, I have noticed many little things in the way animals are worked and handled that could be improved. Poverty and local culture influence the way the oxen are used, but the animals should be yoked comfortably to avoid sores and injuries. Training could also be improved. Such improvements in developing countries means someone has to encourage such change. My hope is that the ideas I have documented here will add a fresh perspective to the many works dedicated to the use of draft animals exclusively in developing nations.

The system of using and training oxen as seen in the United States is not beyond the means of the rural poor across the globe. Yokes, whips, and other tools that are used to work the ox are still locally manufactured with hand tools and technologies that are readily available everywhere. Techniques for training oxen are simple, straightforward, and easily adapted to other languages or cultures. The greatest challenge is sharing new ideas so as to encourage others to try them.

Introducing oxen into cultures that have not used cattle for work is easy where animals are readily available; teaching the importance of animal comfort and the use of improved yoking methods is often more difficult.

Tradition and Example

The New England tradition of driving oxen was imported from Europe and perfected over hundreds of years. Oxen were used for agriculture, logging, and transportation on many small farms in New England as late as the 1950s. They were usually the first animals on America's colonial farms. They opened the way for agricultural development by helping to clear forests, cross the plains, plow tough prairie sods, and transport goods and people across North America.

As is true in other nations today, oxen in early America were the only animals appropriate for the job. They survived where other beasts would have perished. Working in their simple yokes, they accomplished every task necessary to build a nation out of the wilderness.

Training and yoking techniques used in the United States have been passed along from generation to generation, as they continue to be passed along today. Most Americans who keep oxen no longer use them exclusively for farm work, but the techniques they have successfully employed to train and work their cattle are the same as those used during the heyday of oxen in the development of the United States.

The technology for using oxen has not become stagnant in the United States as it has in other developed nations. Ox competitions have raised the level and expectations of animal training and use. Neck yokes and head yokes are pitted against one another in the pulling ring. The designs of each yoking system undergo cycles of experimentation, adoption, and use. Ox training and physical conditioning are often conducted as seriously as the physical training of human athletes.

The level of experimentation pushes ox teamsters in New England to do more and strive for better systems of training and harnessing the animals' power. New England teamsters provide a variety of examples of working animals without the injuries and hazards that face both teamsters and oxen in developing nations.

Challenges and Creativity

Agricultural development arises from the ability to devise a "better" system of agricultural production. For many rural poor the world over, human labor is a major constraint to greater agricultural production. Plowing, weeding, and harvesting are activities for which oxen can easily fill the gap of labor needed. Greater efficiency and timeliness are easily accomplished when oxen are employed in agricultural operations.

Many farmers plow with oxen and then leave them idle for the remainder of the year. Employing the animals in labor-saving and profitable ways takes creativity, which in turn can make greater harvests possible. Acquiring the implements needed for plowing, weeding, and transportation may be a larger constraint than acquiring, training, and employing the animals.

Early American pioneers had to make or find appropriate farm implements. This same struggle exists today in many developing nations. Early America had little infrastructure of roads, communication, or marketing. While fighting wars against foreign nations, engaging in skirmishes with the natives, and weathering ups and downs in the market, the pioneers persisted. Most farmers were left to their own devices to grow successful crops. Often an ax, a plow, a simple sled, and a team of oxen were all they had to begin their farms.

Poor farmers around the world face daily challenges no less daunting. Many today are faced with far greater challenges. Tropical diseases, armed revolutions, shrinking agricultural lands, swelling human populations, and leaders who do not lead, all affect the ability of farmers struggling to produce meager crops against insurmountable odds. For them the ox continues to be of great economic and personal value.

Oxen are not the answer to all agricultural problems nor are they appropriate for all farmers. All users and potential users of oxen must understand both the limitations and the potential of draft animal power. Cattle are a burden. They require feed, water, and security from theft, large predators, and weather extremes. Buying an ox represents a substantial investment for a poor farmer. To lose an animal to disease or theft is a tremendous financial loss. Many farmers prefer hand labor to risking their few resources on a technology they do not understand.

If acreage is increased because animals are employed for plowing and planting, a severe labor bottleneck may occur during weeding. Increasing acreage is of little value if neither additional labor nor the employment of animals is added for weeding and harvesting. Oxen should generate a profit, not create more problems.

Introducing Oxen

Oxen are appropriate where people are genuinely committed to using them. Without strong educational, moral, and technical support, cultures unfamiliar with cattle fail in training and using the animals. Even with such support, local capacity to maintain the technology must be encouraged from the beginning. People must be motivated to help each other train, work, and use their animals. Simply introducing this exciting "new" technology without follow-up inevitably results in failure.

Most successful technologies are passed down from generation to generation. The first time any new technology is introduced, its adoption is slow. Cultural bias, a history without draft animals, and the added burden of caring for livestock may be enough to halt the spread of draft animal use.

Farmers who already use oxen may benefit from improvements in animal training and yoking.

Cattle can be a drain on the resources of a small farm where grazing land is limited or money is lacking for veterinary supplies. Oxen must be readily available and affordable. If they must be trucked in from long distances without ample replacements, their use in the long term may be unsuccessful.

Before the use of oxen is encouraged, an understanding of cattle must come first. Understanding the nature of cattle is the first step, before investing in expensive equipment, elaborate harnesses, and farm implements. All potential users of oxen must realize the animals' limitations and the commitment necessary to make the technology work. The use of oxen can offer an increase in a farm's productivity, but it requires a serious commitment to working the animals.

Training or using oxen involves no magic. Whether oxen are trained in Maine or Mozambique, the commitment to the animals, their training, and their comfort must be top priorities. The importation of animals and fancy equipment is unnecessary. If the local people have the capacity to use their indigenous breeds for oxen and have the interest in training them, the oxen will do the job. If local artisans are taught and motivated to manufacture low-cost yokes with the animals' comfort in mind, the technology will spread.

Farmers who already use oxen may be the first to resist new ideas about improvements to training and yoking. Once they see the advantages of comfortable yokes and well-trained animals, they may be the first to quietly implement improved techniques. Few farmers will pass up an opportunity to learn how oxen can accomplish a variety of tasks on the farm in less time than hand labor.

BREED SELECTION

Local breeds that are carefully selected, then conditioned and properly fed, often achieve what may or may not be achieved by importing exotic breeds and creating an artificial environment. Imported animals or "exotics" may fail miserably.

The breeds or varieties of cattle around the world have been developed over many centuries in specific environments. Although genetic improvements can always be made in any livestock population, don't be too quick to dismiss a breed with which you are unfamiliar. The European breeds or other "humpless" breeds, for example, are often dismissed as work animals in certain African countries, where animals without humps are considered useless and unfit to work. Likewise the smaller local breeds might be dismissed by "outside" experts who have little knowledge of the local environment.

Although a particular breed might not be well suited to a specific local environment, there may be other reasons they are considered unfit to work. For example, certain types of cattle may not be preferred because they have no humps on their backs like the Zebu. In East Africa, cattle without humps cannot work as effectively with withers yokes, as they have been designed specifically for Zebu with large humps. European or other humpless breeds might be a better choice for other reasons, but because of the type of yoke used, animals without humps might be unfairly considered worthless. The imported breeds may also be more prone to local diseases or less adapted to feeding systems with limited water.

TRAINING CONSIDERATIONS

Ox training should be a sequence of events designed to produce animals that are trustworthy in the yoke and willing and able to do the work at hand. Because of local customs, not all oxen may be trained as young animals. Training older animals takes longer and is more complicated than training calves. Rushing through training and beating the animals rarely produces oxen that are eager to work.

A better method is to use a simple series of training exercises, beginning with handling the animals, and progressing to leading them and teaching them to start and stop. Then introduce the yoke and, over a period of weeks, slowly built up the oxen to heavy pulling or plowing. Training is made easier by yoking a young team behind an experienced team. While this step-by-step training may seem to some farmers to be a waste of time, the result is willing animals that are easy to work compared to animals that have been beaten into submission.

Training mature oxen to plow in one day might be an amazing feat that allows an "expert" to visit more villages in a given amount of time, and it can work if the oxen have had previous handling. But the next time they are put into the yoke their attitudes may be altogether different, especially if they are sore from plowing on the first day. In addition, if the animals have to be beaten so severely that they refuse to work or they lie down in the furrow, the new adopters of oxen may not be encouraged to try this new technology.

OX YOKES

When viewing oxen around the world the temptation arises to criticize their yokes. No matter what type of yoke is used, the animals should never suffer due to poor design or fit. An extra few hours spent carving a comfortable yoke pays off in preventing animal breakdowns and injuries.

Every teamster must understand that animals with sores on their necks need an improved yoke design. Yokes that are designed so poorly that they make the animals constantly uncomfortable result not from poverty but from ignorance. Even the poorest person will go without shoes before wearing poorly-fitting shoes that cause wounds. Just as the athlete cannot run in shoes that do not fit, the ox should not wear a yoke that wounds him while he works.

Improvements like carving the yoke beam smooth and increasing its surface area to better fit the animal's neck or withers will greatly increase animal comfort and prevent sores. A good yoke's strength not only protects the animal from injury, but also saves time during plowing season and other critical periods.

A head yoke must comfortably fit the animal's head and horns and be strapped in a manner that does not injure the animal. Wide straps work better than ropes, and a head pad on the forehead keeps the straps from chafing. The yoke should be designed to allow the animals to carry their heads at a level slightly lower than their backs. Animals with their heads held up high or down near the ground are uncomfortable and cannot perform properly in yoke.

Example of the Maasai

Many Maasai farmers in northern Tanzania are just beginning to adopt oxen and a more sedentary lifestyle. These people have traditionally been cattle herders. Growing crops and training oxen are relatively new to them. These poor but proud African farmers have nothing more than crude yokes, simple walking plows, and sleds, yet they do marvelous jobs with their oxen. The key to their success is that they know cattle and how to handle and work them. Their animals are well-fed and -cared for. Ox teams often have special grazing areas near the family compound. They are trained to do whatever work is asked of them, and they are physically conditioned for heavy work before going to the field.

The animals respond to verbal and visual cues, and the teamsters know what to expect from their animals. The result is beautifully plowed fields, no fussing or fighting with the animals, and farmers who are delighted with their work. The oxen wear simple rope nose baskets while weeding. Harvesting is accomplished using a sled and human labor. The technology is not modern, but it represents the beginning of a successful relationship between farmers and oxen.

Implements like a cultivator and cart may be added to the Maasai's collection of tools, but for now the farmers are successful. Their crops are planted on time, they manage the weeding with animals, and harvest is much easier using animal power.

Compared to the option of hiring tractors for plowing, oxen are more affordable and represent a better long-term investment. After working for 4 to 5 years, the oxen are sold for beef at age 8 while in their prime. The farmers start young teams with the proceeds and recoup their initial investment with a little profit to boot.

The Maasai's story highlights a group of farmers who had their animal traction priorities in order. They focused their efforts on the animals and the crops they were growing. They learned how to use oxen by watching other tribes using them. They didn't need outside assistance and had never received training from extension officers or other development organizations. They were not persuaded to purchase fancy implements that sit idle for much of the year.

These farmers pride themselves on their cattle and their crops. Their only concern is the shortfalls of the yoking system they have adopted from neighboring tribes. They are interested in experimenting with new designs that do not cause wounds on their cattle's necks. If all farmers using draft animal power could be so motivated, draft animal power would spread easily to new areas.

Traditionally nomadic cattle herders, the Maasai are beginning to adopt oxen as they learn to grow crops.

Educational Approaches

Potential pitfalls and alternative agendas exist with any method of introducing oxen power for agriculture. The best approach is one that uses a variety of techniques over a long period of time and takes a coordinated effort. Employing a variety of techniques across a region may also be useful for testing the impact of different forms of education. The following are some of the many approaches to introducing oxen and improving their use to farmers in developing nations.

AGRICULTURAL EXTENSION

Some ox training or animal traction programs introduce the technology through the use of extension officers who are familiar with the local people, their customs, and their crops. Extension officers save time because they understand the local language and culture. They must understand the technology and be willing to demonstrate their expertise.

The impact of extension education is short-lived without continuous support and financial aid, which may be a disadvantage if farmers rely solely on extension officers. Sometimes development agencies or animal training experts work with extension officers to help provide longer-term commitment to projects. Most extension officers are responsible for a large area with many farmers. Technical assistance and financial support can help them do a better job of introducing a new technology.

FARMER TO FARMER

Men are most likely to participate in programs that focus on animal traction. As women and children are exposed to the technology, the adoption and use of oxen in their area generally has a greater impact.

Since communication in rural communities can be a serious problem, farmers learn best by example. The best form of education is one that has been identified by the farmers themselves as necessary and in which they can actively participate at every level. This "bottom-up" approach may be difficult to administer and maintain, but the farmers will benefit greatly in the long term.

Using farmer support groups and successful farmers as models and mentors requires the organizational skills of and effort by the person introducing the technology. This approach works best if the organizer spends a long time in an area and wants to have an impact on the local farm community. Working directly with farmers to generate ideas and address their concerns is a great way to share and transfer information.

Some successful animal traction programs are initiated by farmers and run through the use of cooperatives or farmers' groups. These cooperatives and groups are supported by the farmers, the government, foreign development agencies, or companies that may benefit from greater harvests by purchasing crops or selling inputs such as seed, fertilizer, and agricultural chemicals.

Ox training and animal traction training programs, such as this meeting in Uganda, are popular in many areas where oxen and other draft animals have not previously been used.

With funding, support, and influence from outside the local community, some agencies or companies may have specific concerns or desires. These must be understood and addressed beforehand, especially in countries without strong agricultural leadership.

VOCATIONAL SCHOOLS

Vocational schools provide hands-on experience and help vocational students learn alternative ways of farming. Many vocational schools have fields and plots where the students work. Students are usually eager to try anything that will ease the manual labor they must do in the field. In vocational schools, high standards of animal training and use can be achieved because students work for months to perfect their ox yokes, training techniques, and the oxen they use. While vocational students are usually willing to learn new techniques, they sometimes become frustrated when they return home and find no support for their new ideas.

Agricultural Shows

Agricultural shows at a national or regional level provide a good place to see how area farmers react to the idea of ox power by providing hands-on demonstrations and discussions. Such shows also challenge farmers to view new systems of training and yoking oxen. Large corporations that sell products or merchandise to farmers may be enticed to support either the entire show or training programs, demonstrations, and other events at the show.

LOCAL COMPETITIONS

Competitions where farmers are invited to bring their ox teams for plowing contests, ox pulling, and ox cart competitions generate interest and challenge teamsters to improve their techniques. Farmers are usually persuaded to come when prizes are offered. Farmers who do not bring their own teams may learn by watching others and seeing their successes or failures. When a successful event spawns subsequent events, changes and new ideas follow. Historically this has been done in North America, and even with Cuba's recent reintroduction to oxen,

as mandated by the government, ox competitions have flourished as a way to spread the technology and encourage innovations.

RESEARCH INSTITUTIONS

Research institutions often provide sites for testing new ideas and techniques in a controlled environment. Demonstration plots or farms may be used as tools to demonstrate what may be accomplished under ideal conditions. While such demonstrations may generate interest among farmers, the impact will be greater if the demonstrations are located in local villages or other visible areas. The animals and the crops at research institutions are also often in better condition than would be found in the villages, and farmers may point out that the labor and input does not reflect the reality on their farms.

MEDIA

Television, radio, and local billboards may be used to introduce technology and help farmers become interested in using oxen. These media may not directly address the needs and concerns of the farmers, but instead may provide the impetus for making informed decisions before farmers decide to use oxen in their operations.

Many cultures have had cattle for decades but never realized the potential of using the animals for draft power. Ox power technology may be introduced in many ways. Forcing or persuading people to use a technology that is not appropriate is doomed to failure. The key to success is to first interest farmers in using oxen.

CHAPTER ROUND-UP

Oxen are not the answer for all people who are trying to make improvements in their agricultural systems. Some cultures have an aversion to keeping animals or have geographic and biological concerns that may make keeping cattle difficult or risky. For those who wish to improve their ability to plant, harvest, and transport crops and forest products, however, oxen may be ideal. ★

Paul Starkey
Oxgate, Reading, England

I first met Paul Starkey, from Reading, England, when he visited the United States in 1991. He said he was a consultant who worked around the world with oxen and wanted to meet with me. I was delighted to make the acquaintance.

Over the course of his two-day visit, I was his guide to numerous New England fairs, including Fryeburg Fair in Maine. At the time we engaged in the most thought-provoking dialogue I had ever had on oxen.

His experience included traveling to more than 100 countries around the world, where his work with animal traction (draft animal power) focused largely on the use of oxen. At the time, I remember realizing how little I knew about oxen, outside of my focused perspective on using them in living history farms or in competition at agricultural fairs.

Paul's visit inspired me to learn more about oxen outside the United States. His photos and ideas mesmerized me and inspired me later to travel to Africa. I must admit, look-ing back, that my meeting with Paul largely inspired my Ph.D. research on the use of oxen by the Maasai people in Tanzania, who historically are pastoralists, and who only recently adopted the use of oxen in growing agricultural crops.

As I did my preliminary research, Paul's many books, conference proceedings, and articles became critical to understanding the use of oxen by the world's poor. When I contacted Paul in 1996, before traveling to Tanzania to begin my field work, he also became a resource, helping me make connections with people doing research and extension work in East Africa.

Since that first meeting, Paul has continued to stay in touch annually with his unique photo Christmas cards portraying oxen at work in different countries that he has visited. We met in 2002 in Uganda at a conference called *Modernising Agriculture: Visions and Technologies for Animal Traction and Conservation Agriculture*. We then traveled together to meet some farmers and extension people doing great work with oxen in Uganda.

We met again in 2004, when I visited England and Paul was my host, to present a paper on *Ox Yokes: Culture, Comfort and Animal Welfare* at a conference in England (see: www.taws.org/TAWSworkshop 2004.htm). It was exciting to meet his family, who were all global travelers, and see his amazing collection of ox yokes, equipment, and books on oxen. A Christmas card showing oxen in Cuba and our discussion about that country that spring inspired me to travel there later in 2004.

Paul Starkey continues to be an inspiration to me with his work overseas related to animal traction. His travels, tales, experiences, and adventures related to oxen probably make him the most experienced person globally with regard to using oxen. He is certainly the most prolific writer on the subject, and most of his work can be found on his Web site: www.animaltraction.com.

Overview of Ox Breeds

	Comparing Ox Breeds				
BREED	**Color**	**Weight***	**Height****	**Tractability*****	
Ayrshire	brownish, red and white	2,000 lbs. (4,400 kg.)	medium	7	
Brown Swiss	light to dark brown	2,400 lbs. (5,280 kg.)	large	2	
Charolais	white	2,300 lbs. (5,060 kg.)	large	6	
Chianina	white	3,000 lbs. (6,600 kg.)	huge	10	
Devon	red	1,600–1,800 lbs. (3,520–3,960 kg.)	medium	10	
Dexter	black, dark red, dun	1,000 lbs. (2,200 kg.)	small	9	
Dutch Belted	white belt around black body	2,000 lbs. (4,400 kg.)	medium	3	
Guernsey	brown or fawn with white patches	1,800 lbs. (3,960 kg.)	medium	1	
Hereford	dark red with white face	2,200 lbs. (4,840 kg.)	medium	3	
Holstein	black and white or red and white	2,500 lbs. (5,500 kg.)	large	3	
Jersey	any shade of brown	1,600 lbs. (3,520 kg.)	medium	8	
Lineback	black with white stripe down back and belly	2,000 lbs. (4,400 kg.)	medium	7	
Milking Shorthorn	red, red and white, roan	2,300 lbs. (5,060 kg.)	large	4	
Scotch Highland	all black, brown, gray, or white	2,000 lbs. (4,400 kg.)	medium	9	
Zebu (American)	gray or grayish brown	2,000 lbs. (4,400 kg.)	large	8	
Texas Longhorn	any color or combination	1,800–2,000 lbs. (3,520–4,400 kg.)	medium	10	
Chianina × Holstein	black or brownish black	3,000 lbs. (6,600 kg.)	huge	7	
Devon × Holstein	solid black	2,000 lbs. (4,400 kg.)	medium-large	7	
Hereford × Holstein	black and white with white face and head	2,200 lbs. (4,400 kg.)	large	2	
Jersey × Holstein	solid black or solid black-dark brown	1,800 lbs. (3,960 kg.)	medium	5	
Shorthorn × Holstein	black, black and white, blue roan	2,300 lbs. (5,060 kg.)	large	4	

Yoke and Bow Sizes

Temperament	Other
active and alert	attractive long horns
docile and slow moving	grows rapidly
moderately alert and active	heavily muscled
alert and excitable, at times to a fault	grows rapidly
quick and alert	easy to color match
quick and alert, at times to a fault	long horns, heavy muscling
moderately docile	rare breed
easy-going	lean and angular
moderately docile	rare breed
docile	grows rapidly
moderately active and excitable	easy to color-match
moderately active and alert	rare breed
moderately docile	matures slowly
active	long shaggy hair
active	hump on back
active and alert	impressive horns
moderately active	
moderately quick and alert	easy to color-match
easy-going	
moderately active	easy to color-match; becoming more common on dairy farms
moderately docile	fatten easily

Rough Beam Sizes for Neck Yokes

Yoke Size	Beam Dimensions (inches) width x depth x length
4	4 × 4 × 28
5	5 × 5 × 35
6	6 × 6 × 42
7	6.5 × 7 × 49
8	7 × 8 × 56
9	7.5 × 9 × 63
10	8 × 10 × 70
11	9 × 11 × 70+
12	10 × 11 × 70+

Stock Sizes for Bows

Yoke Size	Bow Dimensions (inches)
4	1 × 1 × 32
5	1.25 × 1.25 × 40
5	1.5 × 1.5 × 48
6	1.5 × 1.5 × 56
7	1.5 × 1.5 × 62
8	1.75 × 1.75 × 64
9	1.75 × 1.75 × 66
10	2 × 2 × 68
11	2 × 2 × 70
12	2 × 2 × 72

* Average weight of a mature ox.
** Comparative height; when taken together with average weight gives you an idea of the breed's mass
*** On a scale of 1 to 10: 1 being the easiest and 10 the most challenging for an inexperience teamster to train and work in a yoke.

Calculating Proper Nutrition

Pearson's Square

This system allows you to calculate the correct proportion of two feeds to meet an animal's protein or energy needs. Although outdated for dairy and beef cattle on complex diets, Pearson's Square is a simple method that works well for steers or oxen. Here's how it works.

Nutrient Requirements of Steers/Oxen				
Body weight lbs.	**Gain/day lbs.**	**DMI* lbs.**	**Protein %**	**Energy (TDN)** %
2,200***	0	33.6	6.8	48.4
1,800***	0	28.9	6.8	48.4
1,800***	0.5	30.9	7.0	52.0
1,500***	0	25.2	7.0	48.4
1,500***	1.5	29.0	7.6	59.7
1,300***	1.0	25.4	7.6	55.8
1,000	0.5	20.2	7.5	52.5
1,000	1.5	22.5	8.2	59.5
1,000	3.0	23.6	9.3	72.0
800	0.5	17.1	7.7	52.5
800	1.5	19.0	8.8	59.5
800	3.0	19.9	10.4	72.0
600	0.5	13.2	8.2	52.5
600	1.5	15.3	9.7	59.5
600	3.0	16.1	12.0	72.0
400	0.5	10.1	8.9	52.5
400	1.5	11.3	11.4	59.5
400	3.0	11.9	15.2	72.0
300	0.5	8.2	9.5	52.5
300	1.5	9.1	12.9	59.5
300	3.0	9.6	18.0	72.0

*Total dry matter intake.   **Total digestible nutrients.
***Adapted from M.E. Ensminger's *Beef Cattle Science* based on large-frame beef steers and mature bulls.

1. Choose two feeds: one grain and one forage.
For example:

Name of Feed	CP %	TDN % (or Mcal/lb)
grain = corn	8.9%	92%
hay = alfalfa	16.0%	57%

2. Use the Nutrient Requirements of Steers/Oxen table (opposite) to find each of your steer's requirements for CP and TDN based on his body weight and the desired average daily gain (ADG). See chart below for example:

	Live Weight	ADG	DMI	CP%	TDN%
Steer #1	400 lbs.	3 lbs./day	11.9 lbs./day	15.2%	72%
Steer #2	400 lbs.	0.5 lbs./day	10.1 lbs./day	8.9%	52.5%

3. Set up the Pearson's Square. Begin by drawing a square on a piece of paper. To determine CP, at the upper left corner write down the amount of CP in grain, in this example corn at 8.9 percent. At the lower left corner, write the amount of protein in the hay, in this example alfalfa at 16 percent. In the center of the square, write the amount of CP you want to end up with for that steer, in this example 15.2 percent.

% in grain = _____   grain needed = _____ / Total = _____%

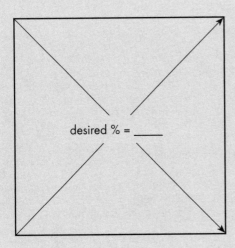

desired % = _____

% in hay = _____   hay needed = _____ / Total = _____%

_____ / Total = 100%

4. Moving from the upper left toward the lower right (following the arrow in the illustration), subtract the smaller number from the larger number. Write the answer (resulting in a whole positive number) in the lower right-hand corner. Moving from the lower left toward the upper right (again following the arrow), subtract the smaller number from the larger number. Write the answer (again resulting in a whole positive number) in the upper right-hand corner.

The number at the upper right tells you how many pounds of grain, in this example 0.8 pounds of corn, you need to combine with the amount of hay, in this example 6.3 pounds of alfalfa shown at the lower right, in order to balance the ration the way you want it. Combining these two feedstuffs will give you the right proportions, but not the right amount of daily feed for steer #1.

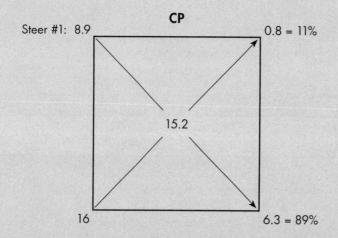

5. To convert these amounts into percentages, add them together and divide each by the total. In our example the diet would consist of 11 percent corn and 89 percent alfalfa hay.

6. Now that you have calculated the correct percentage of each feed ingredient based on crude protein, check the feed mix to see if it has the right energy level for steer #1:

% grain in diet × TDN in grain = energy from grain
(11%)   (92%)   (10.12%)

% hay in diet × TDN in hay = energy from hay
(89%)   (57%)   (50.73%)

Total energy = (~61%)

7. Now compare these two sets of figures. Will the two feeds you have chosen combine to meet the steer's nutritional needs in terms of both CP and TDN?

8. If the answer is yes, you are done. You have selected a simple ration that meets the needs of your steer at the desired rate of growth.

9. If the answer is no, you will have to either select other feeds or not feed your animal according to your original plans. If the energy is too low, you may have to feed a grain with more energy, but make sure the protein is at the desired level. If the energy is too high, you may have to feed less grain and more alfalfa hay. Or you might choose two other ingredients and start your calculations again.

In our example, Pearson's Square suggests that 11 percent corn grain mixed with 89 percent alfalfa hay meets both the energy and protein needs of steer #1.

10. The last thing you need to do is convert the percentage of each feed in the diet to pounds of dry matter feed. In our example for steer #1:

grain:
% in diet × lbs. DMI for steer = required lbs. DMI
(11%)   (11.9 lbs.)   (1.3 lbs. corn)

hay:
% in diet × lbs. DMI for steer = required lbs. DMI
(89%)   (11.9 lbs.)   (10.6 lbs. alfalfa hay)

Most feeds are analyzed as dry feeds without any moisture in them, so you may have to do one more calculation to convert from dry matter to as fed. Dry matter is always a lower number than the pounds of moister feed the animal actually receives. To convert from dry matter to as fed, divide pounds DMI by the feedstuff's percent dry matter (feed with all the water taken out).

In our example, from the Nutrient Composition of Feeds Commonly Fed to Oxen table we see that corn has 87 percent dry matter. The steer's diet calls for 1.4 DMI pounds of corn, divided by 87 percent gives you 1.6 pounds as fed. The steer's diet calls for 10.5 DMI pounds of alfalfa hay, divided by 91 percent dry matter gives you 11.9 pounds as fed.

Steer #2 is supposed to grow more slowly with only 0.5 pounds of gain per day, a rate that would be difficult to achieve on a diet combining corn with alfalfa. Pearson's Square suggests a diet of all corn. While such a diet might meet the animal's protein requirements, it overfeeds energy. On a corn diet the steer would become ill and may die due to lack of fiber in his diet. Too much energy and too little fiber is a deadly combination.

With steer #2 you must either feed corn with alfalfa and accept the idea that the steer will grow faster than 0.5 pounds per day, or choose different feeds. You can likely find a better combination of feeds, or perhaps one forage that alone will provide all the animal's requirements.

Pearson's Square does not replace common sense. If it suggests feeding an inappropriate combination, go back to the drawing board and start with more appropriate ingredients.

This system works well provided your chosen feeds are at least as good as the average book values or your feeds have been analyzed for these nutrients. The success of this system also depends on an accurate evaluation of each animal's weight, age, and breed or frame size.

Nutrient Composition of Feeds Commonly Fed to Oxen

All CP and TDN values are listed on a dry matter or moisture-free basis.

Feedstuff	Dry Matter %	%CP*	TDN%**
Alfalfa			
prebloom pasture	20	21.1	61
full bloom pasture	24	19.2	55
early cut hay	91	19.8	58
late cut hay	91	16.7	61
early cut silage	35	16.7	52
late cut silage	36	14.9	59
Bakery Waste*	91	11.1	89
Barley, grain	88	13.2	85
Beet Pulp, dehydrated	91	9.7	74
Birdsfoot Trefoil			
pasture	19	19.3	68
hay	91	15.3	59
Brewer's Grain, dehydrated	92	30	71
Bromegrass			
early pasture	34	17.1	74
early cut hay	88	10.5	58
late cut hay	92	6.6	51
Canarygrass, Reed			
pasture	23	17	61
hay	89	10.2	49
Clover, Red			
early pasture	20	19.4	69
late bloom pasture	26	14.6	64
hay	88	14.7	59
Clover, White			
pasture	18	24.7	76
hay	89	22.4	65
Corn			
grain	87	8.9	92
silage	26	8.3	68

*CP = crude protein **TDN = Total digestible nutrients *** = Nutrient content varies tremendously.
Adapted from M.E. Ensminger's *Beef Cattle Science*.

Nutrient Composition of Feeds Commonly Fed to Oxen *continued*

All CP and TDN values are listed on a dry matter or moisture-free basis.

Feedstuff	Dry Matter %	%CP*	TDN%**
Fescue, Meadow Hay	89	8.1	54
Ken. Bluegrass			
early pasture	31	17.4	72
early cut hay	89	10.2	61
late cut hay	89	6.3	46
Oats			
grain	89	13.3	77
silage (dough stage)	35	10.0	57
straw	92	4.4	50
Orchardgrass			
early pasture	23	24.5	67
midbloom pasture	27	10.1	60
hay	89	10.5	57
Soybean			
meal, 44%	89	49.8	85
whole roasted	92	42.0	92
Sorghum			
grain	89	10.5	83
silage (mature)	32	8.6	53
Timothy			
early pasture	26	12.5	64
midbloom pasture	29	9.1	64
early cut hay	89	10.7	58
late cut hay	89	7.7	58
silage	34	10.5	59
Urea			
45% nitrogen	99	285.0	0
Wheat			
grain	89	14.7	87
bran	89	17.5	70
middlings	89	18.5	83
straw	90	3.6	44

Glossary of Ox Terms

Ard. A crude plow used in many nations for soil preparation prior to planting. Most often it is simly a large forked stick, sometimes fitted with a metal point.

Bell ox. The lead ox in the herd that wore a bell when turned out to pasture.

Big wheels. A pair of wheels, taller than the team of oxen, which straddled the log, lifting one end. These big wheels were used for transporting logs. They were commonly used in forests that were in flat regions like the upper Midwest.

Bitch link. A pear shaped link on a short chain, which was often used as a grab hook to hitch teams together or hitch them to a load. It is also sometimes referred to as the callabash ring.

Bird's-eye tenderloin. Ox meat showing marks of a teamster's goad (nails in the end of their long driving sticks).

Bob, bob sled. A two-runner sled used for hauling logs out of the woods. Also called a lizard or yarding sled.

Breeching. See Britchen.

Bridle chain. A chain wrapped around the front sled runners to serve as a brake when going down hills.

Britchen. A harness strap around an ox's hindquarters to help hold back the load while going downhill. Also called Breeching and Brichen.

Bull puncher, bull whacker, bull skinner. All terms used to describe the ox teamster.

Bullock. A term to describe a steer or ox used by other English speaking cultures.

Bow bunch. Swelling caused by chafing of bow.

By the bushel, by the inch, by the mile or by the piece. Ways of describing logging done for contract.

Cattle power. Ox power, sometimes referred to as bull power.

Calabash ring. Same as a Bitch Link, although this term is often used to describe the pear shaped ring on the yoke, that the chain attached to the implement or load hitches to.

Calf hutch. A simple three-sided structure designed to house an individual calf until they are weaned.

Cannula. A plastic or synthetic material covering an opening into some body part of the bovine. Most often the rumen or duodenum with a cap or cover that allows researchers to take samples from the gut in a living and working animal.

Clevis. A C-shaped piece of hardware with a removable bolt running through it. A small clevis is sometimes used to repair chains or connect two short chains together. It is often used to provide a place to hitch the chain to a sled or other implement, when using oxen.

Colic. Impaction of feed in the gut.

Colostrum. The first milk from a cow after she has given birth.

Come-along. This is a winch type device used to tighten loads or move large objects. Many modern come-alongs are designed to used a hand crank. In the past, the use of pulleys and oxen could be used to move large objects much like a modern come-along.

Come here, come haw. To turn left or come toward the ox teamster.

Crooked stick. This is another old term for an ox yoke.

Dog. A steel spike attached to a sled runner that digs into the ground to prevent it from sliding backwards when going up hill.

Dolly wheels. When pulling something like an animal-drawn mowing machine, a set of small wheels under the tongue that helps to carry the weight of the machine, reducing fatigue on the animals.

Drench. Liquid medication given to the animal through the mouth or esophagus using a tube or a large dose syringe.

Dung sniffer. A term sometimes used for an ox teamster.

Evener. The evener was something used primarily by large horse hitches as a way to evenly distribute the load among many horses. Most oxen were hitched in tandem with a single chain, so an evener was not required. If oxen are hitched abreast the evener would be required. The vertical evener was developed to aid in adjusting the load pulled by multiple teams in tandem.

Gee, Gee off. Gee is the most common command used to direct an ox or oxen to turn right. Gee off is used to mean turn away from the teamster. Back Gee is sometimes used to make the off steer back up a few steps while the nigh steer makes a tight right turn.

Girt. An ox was measured by its girt or girth when sale prices for oxen were agreed upon. The heart girth or girth was also used in pulling contests to make sure oxen of similar sizes competed.

Goad. This was a carved stick used to drive oxen. Its length varied depending on the size of the oxen, and the teamster.

Most often it was 4–5 feet long, and carved of flexible hardwood species. A goad stick implies in New England that there is no lash attached to it.

Gore stick steer. When an ox broke its leg or was injured in the forest while logging, it was butchered and eaten by the loggers.

Grab hook. A hook at the end of a chain that can be hooked or "grab" into the chain. It is useful in making adjustments to chain length.

Grab ring. A ring similar to the callabash ring or bitch link.

Handy. A term used to describe how well trained a team of oxen are.

Haw. The most common command used to direct an ox or oxen to turn left.

Haylage. An ensiled form of grass and legumes. The fresh forage is dried down slightly in the field, and then placed in a silo, packed in tight, and sealed from the air. The result is an ensiled feed that is both palatable and nutritious to cattle.

Haymow. Another term for a hayloft, where loose hay is stored

Headland. A strip of unplowed land at the end of furrows, usually near a fence or field border.

Hovel. A building or shed for housing oxen in semi-permanent logging camps.

Horn knobs (also called ox balls). Brass covers for the tips of the horns added to prevent oxen from goring each other. They also used for decoration in animals that are exhibited at fairs and events.

Interfere. When an ox's hoof or foot interferes with his other hoof, or the hoof of a nearby ox.

Leaders. The oxen in front when ox teams are hitched together in tandem, as they often were in the forest, to pull large logs or during long hauls. They were usually the team that would better respond to the teamster's demands.

Log boat. A long sled used for skidding logs in wet places.

Lizard. A sled for skidding logs, usually only used for lifting the front of the log. Often made from a Y-shaped crotch of a tree. Also called crazy drag, go-devil, snow snake or travois.

Lunge space. 2 feet of manger space, which allows cattle to lie down in a normal position, and to throw their heads forward as they get up.

Mating a team. Putting together or breeding cows to bear calves that will grow, develop, and act alike.

Nigh or near ox. The ox nearest the teamster. If standing behind a pair of oxen, this is the ox on the left. It is the ox to the right of the teamster, when walking beside the team.

Off or far ox. The ox in a pair that is furthest from the teamster. The off ox, is the ox on the right when standing behind a pair of oxen. When the teamster is standing or walking beside the team, the off ox is on his right.

Ox. A castrated bull that has reached four years of age and has been trained to work. Historically, a steer was not considered old enough for real work until he was four years old, thus the term was used to differentiate between a young and not fully grown animal and one that was at least near maturity. Oxen is plural for ox.

Ox balls. Metal (often brass) balls that are threaded and screwed in place on the tips of an ox's horns.

Ox goad. A long stick with a small brad or nail near the end used to persuade reluctant oxen to work in the yoke. Also called goad stick or goad. In New England the goad is about 4 feet long, ¾ to 1 inch thick at the handle and about ½ inch thick at the small end.

Ox harness. Oxen harness similar to those used on horses, however if the horse collar was used it was often flipped upside down to raise the hitch point above the prominent shoulders of the ox. A three pad collar was designed specifically for the ox, based on the anatomy of their shoulders and neck.

Ox muzzle. A basket made of wire, leather, wood or other materials that is placed on the ox to prevent him from grazing or trying to eat while in the yoke.

Ox prone. A frame used to hold an ox's foot during shoeing.

Pole team. The team of oxen closest to the sled or wagon. They were usually the only team attached to the load by a pole. They would turn the sled or wagon and slow or stop it if necessary. They were sometimes also called wheelers or the butt team.

Polled. Naturally lacking horns. The Angus, Galloway, and Polled Hereford are examples of polled breeds.

Pressure quicking. When nailing on an ox shoe, if the nail is too close to the sensitive tissue, the animals may become lame simply due to the pressure on this sensitive tissue.

Quicking. A term describing when a shoer, nailing on an ox shoe, drives the nail into the sensitive tissue, causing the animal to bleed and become lame. Oxen can also get an infection from the debris or bacteria that might have been carried into the flesh when the nail went in.

Racking a load or sled. Breaking a frozen sled or log loose by jamming the oxen to the right and left until the sled is freed. It is often easier to simply use a pole to pry the sled upward to break the sled free of the ice on the runners.

Roan. A coat color made up of alternating red and white hairs.

Running W. A set of ropes and pulleys that are placed on an unruly animal. Historically animals (usually horses) that are prone to running away were subjected to this form of restraint and training. As the animal bolted the ropes were pulled and the animals front feet were pulled out from under them.

Scoot. A sled designed with two long solid runners used most often in winter in Northern areas to haul firewood, small logs, maple sap tanks, and other farm or logging supplies.

Scours. Diarrhea.

Shaves. When an ox is hitched single to a cart they require shaves, which attach to either side of the side yoke to hold the front of the cart up.

Shoeing sling, shoeing stocks, or shoeing rack. A strong wooden frame or rack with canvas or leather straps to support the weight of an ox and make shoeing easier. Also called an ox sling, a shoeing stock or ox rack.

Singletree. When using a single ox, like a single horse, the animal will require a singletree. This connects both of the animal's trace chains together behind the animal. It provides a single hitch point behind the animal, to which might be attached a plow, cultivator or other implements.

Skidding. Pulling logs out of the forest, usually on the ground or on a scoot.

Skidding team. A team used to drag logs out of the forest, usually to a landing, where the logs are then loaded onto a truck or larger sled.

Skid road. A road, skid trail, or path on which the oxen pull logs.

Slide yoke, slip yoke. A yoke designed to allow the yoked oxen more freedom of movement than the traditional solid wood ox yoke. It had individual neck pieces for each ox, which slid to the right and left, as the oxen moved. These were maintained an equal distance from the center, by means of hardware to which it was attached.

Slip hook. A type of hook on a chain that would not be used to hook into the chain. It was designed to wrap around a log and tighten by allowing the chain to slip through it, as the oxen leaned into their yoke.

Snow roller. A huge wooden roller drawn by ox teams in order to pack down a road. They were used before snow was plowed from roads for motor vehicles.

Steer. A castrated bull that may or may not be trained to work.

Stump puller. Any of a number of devices used to remove stumps from skid trails or clearings. Stump pullers were designed to use oxen or horses to unscrew stumps or yank out of the ground. Other kinds consisted of a series of pulleys or levers to pull stumps out.

Sweet feeds. Commercially prepared pelleted feed containing molasses.

Team. The term is used loosely and generally means multiple pairs of oxen hitched together although that number may vary between three and six pairs.

Tongue stop. A device that keeps the pole or tongue that the animals are hitched to from running through the yoke. This helps to hold the load, and also allows the yoked oxen to back up a loaded wagon or cart. Most tongue stops are made of steel and bolted onto the tongue of the cart or wagon. Simple tongue stops can be made using the natural crotch of a tree.

Trocar. This devise when used with a small cannula is a last resort means of saving an animal that is dying from bloat. It is last resort, because the animal will often die of peritonitis or an infection of the abdominal cavity after it is used.

Twitching logs. Pulling individual logs out of the forest to a common landing or gathering place where they were then loaded onto sleds or wagons for further transport.

Whiffletree. Another name for an evener, or double tree.

Yoke. The wooden beam with bows and iron hardware that is used to harness the power of the ox. A pair of oxen is also called a "yoke" or a yoke of oxen. To place the yoke on is "to yoke" the oxen.

Metric Conversion	
Imperial	**Metric**
WEIGHT	
1 ounce	28.4 g
1 pound (lb.)	454 g
1000 lb.	454 kg
LENGTH	
¼ inch	.64 cm
½ inch	1.28 cm
1 inch	2.5 cm
1 foot	.3 m
1 mile	1.6 km

Bibliography

American Agriculturist. "Training Steers." Reprinted from an article over 100 years old as "Early Methods of Training Oxen" in *Tillers Report.* 1987 Fall, 8–9.

Ashburner, John and Paul Starkey. *Draught Animal Power Manual.* Rome: Food and Agriculture Organization of the United Nations, 1994.

Balt, Linda E. "Having Fun," *Rural Heritage.* 1994 Summer, 19:3, 40.

Barden, Les. *Training the Teamster.* Farmington, NH: Barden Tree Farm, 1984.

Barnes, Thomas. "Making an Ox Yoke," *Foxfire 2*, ed. Eliot Wigginton. New York: Anchor Press, 112–117, 1973.

Barney, Dwight. *4-H Working Steer Manual.* Durham, NH: New Hampshire Cooperative Extension, 1981.

———. "Working Steers," *Rural Heritage.* 1994 Evener, 19:2, 16.

Blynt, Ruth Ann. "Oxen in Upstate New York," *Rural Heritage.* 1995 Spring, 19:2, 22–23.

Buchanan, Milliard. "Breaking and Training Oxen," *Foxfire 4*, ed. Eliot Wigginton. New York: Anchor Press, 260–268, 1977.

Bunnell, Doug. "Getting Started with Oxen," *Rural Heritage.* 1996 Winter, 21:1, 56–59.

Bunting, William. "Brooks Sproul, Maine Oxman," *Small Farmer's Journal.* 1986 Spring, 10:2, 28–35.

———. "Clyde Robinson, Cagey Maine Oxman," *Small Farmer's Journal.* 1987, 11:4, 33–38.

Campbell, Joseph K. *Dibble Sticks, Donkeys, and Diesels: Machines in Crop Production.* International Rice Institute: Manila, Philippines, 1990.

Cannon, Arthur. *The Bullock Drivers Handbook.* Night Owl Publishers: Shepparton, Australia, 1985.

Clough, William. "Oxen," *Country Journal.* 1976 September, 38–41.

Conroy, Drew. "Advanced Oxen Training." *Tillers Tech-Guide.* Kalamazoo, MI: Tillers International, 1995.

———. *Advanced Training of Oxen.* VHS. Butler Publishing and Tools: LaPorte, CO, 1988.

———. "Answering Hollywood's Call," *Rural Heritage.* 1996 Spring, 21:2, 14–17.

———. "Are Your Oxen Training You?" *Rural Heritage.* 1997 Evener, 22:3, 87–90.

———. "Barn Raising," *Rural Heritage.* 1994 Spring, 19:1, 15–18.

———. *Basic Training of Oxen.* VHS. Butler Publishing and Tools: LaPorte, CO, 1987.

———. "Berry Brook Farm and Ox Supply." *Small Farmer's Journal.* 2005 Fall, 29:4, 43–44.

———. "Bovine Behavior," *Rural Heritage.* 1998 Evener, 23:2, 106–109.

———. "Building Woodpiles and Confidence," *Rural Heritage.* 1995 Autumn, 20:4, 22–24.

———. "Buying a Team of Oxen," *Rural Heritage.* 1999 Summer, 24:4, 64–66.

———. "Calf Training." *Rural Heritage.* 2002 Evener, 27:2, 41–42.

———. "Cuba's Organic Production," *Rural Heritage.* 2005 Holiday, 30:6, 72–75.

———. "The Demise of the Ox?" *Rural Heritage.* 1995 Summer. 20:3, 7–9.

———. "Donkeys and Oxen in Uganda," *Rural Heritage.* 1996 Summer, 21:4, 57–59.

———. "Driving Oxen with Lines," *Rural Heritage.* 1999 Evener, 24:2, 102.

———. "Equines vs. Bovines," *Rural Heritage.* 2001 Winter, 26:1, 59–61.

———. "Feeding Oxen," *Rural Heritage.* 1996 Holiday, 21:6, 26–29.

———. "The Future of Draft Animals on Historic Farms," in Proceedings of the 20th Anniversary Meeting of the Association of Living History Farms and Museums on June 17–24 1990, Vol XIII, 210–214, 1993.

———. "Harnessing Ox Power," *Rural Heritage.* 2000 Winter, 25:1, 30–32.

———. "He Likes Big Oxen," *Rural Heritage.* 1997 Spring, 22:2, 68–70.

———."He Sets The Standard," *Rural Heritage.* 1995 Holiday, 20:5, 11–13.

———."Head Yokes vs. Neck Yokes," *Rural Heritage.* 1998 Holiday, 23:6, 66–69.

———."Heifers and Cows for Work," *Rural Heritage.* 2003 Winter, 28:1, 52–54.

———."Hoof Trimming," *Rural Heritage.* 1997 Summer, 22:4, 66–69.

———."How Much Steers Cost," *Rural Heritage.* 2003 Spring, 28:3, 96–98.

———."How Oxen Learn," *Rural Heritage.* 1998 Spring, 23:3, 40–41.

———."Interpreting Oxen in Early America," *Rural Heritage.* 2004 Autumn, 29:5, 80–83.

———."Judging Oxen," *Rural Heritage.* 2003 Holiday, 28:6, 52–54.

———."Keep Traveling Oxen Healthy," *Rural Heritage.* 2000 Autumn, 25:5, 49–51.

———."Maasai Agriculture and Wildlife Conflicts" in the FAO's Livestock, Environment and Development (LEAD) Initiative's Electronic Newsletter, 2003.

———."Making an Ox Yoke," *Rural Heritage.* 1993 Winter, 18:4, 18–20.

———."Making Bows for an Ox Yoke," *Rural Heritage.* 1996 Autumn, 21:5, 38–41.

———."Making Ox Shoes," *Rural Heritage.* 1999 Winter, 24:1, 59–63.

———."Midwest Ox Drovers' Association," *Rural Heritage.* 1996 Autumn, 21:5, 26–27.

———."Milking Devons," *Rural Heritage.* 1997 Holiday, 22:6, 86–88.

———."Modern Farming with Oxen," *Rural Heritage.* 1998 Autumn, 23:5, 63–65.

———."Multiple Ox Hitches," *Rural Heritage.* 1999 Holiday, 24:6, 51.

———."New England Log Scoot," *Rural Heritage.* 2003 Evener, 28:2, 24–26.

———."New England Ox Events," *Rural Heritage.* 1996 Evener, 21:3, 72–74.

———."9 Year Old Ox Teamster," *Rural Heritage.* 1995 Summer, 20:3, 28–30.

———."No Shortcuts for Training Oxen," *Rural Heritage.* 2000 Summer. 25:4, 26–28.

———."Obstacle Competitions," *Rural Heritage.* 2004 Spring, 29:3, 62–64.

———."Ox Collars and Harness," *Rural Heritage.* 1997 Winter, 22:1, 13–14.

———."Ox Drover's Gathering," *Rural Heritage.* 2001 Spring, 26:2, 55–57.

———."Ox Housing," *Rural Heritage.* 1999 Spring, 24:3, 75–77.

———."Ox Logging," *Rural Heritage.* 1995 Winter, 20:1, 15–18.

———."Ox Mecca (Fryeburg Fair)," *Rural Heritage.* 1999 Autumn, 24:5, 63–65.

———."Ox Muzzles," *Rural Heritage.* 1999 Evener, 24:2, 20–21.

———."Ox Pulling," *Rural Heritage.* 1995 Spring, 20:2, 30–33.

———."Ox Yokes and Comfort," *Rural Heritage.* 2004 Summer, 29:4, 71–74.

———."Oxen in Harness," *Rural Heritage.* 1994 Evener, 19:2, 17.

———."Oxen, Selecting Your First Team," *Small Farmer's Journal.* 1989, 13:3, 44–45.

———."Reshaping Crooked Horns," *Rural Heritage.* 1996 Winter, 21:1, 16–17.

———."Selecting a Team of Oxen," *Rural Heritage.* 1994 Autumn, 19:4, 18–21.

———."Selecting and Teaming Oxen," in *The Tillers TechGuide.* Kalamazoo, MI: Tillers International, 1992.

———."Sharing the Knowledge," *Rural Heritage.* 2002 Winter, 27:1, 25–28.

———."Shoeing Oxen," *Rural Heritage.* 1997 Autumn, 22:5, 70–74.

———."Starting a 4-H Working Steer Club," *Rural Heritage.* 2000 Evener, 25:2, 102–105.

———."Stoneboats for Exercising and Working Oxen," *Rural Heritage.* 2000 Holiday, 25:6, 102–105.

———."The Traditional Ox Team and Its Yoke," *The Tillers Report.* 1988, 8:1, 1–5.

————."Understanding and Training Cattle, Part I," in *Draught Animal News*. Edinburgh: Centre for Tropical Veterinary Medicine, University of Edinburgh, 1997 June, 26, 33–39.

————."Understanding and Training Cattle, Part II," *Draught Animal News*. Edinburgh: Centre for Tropical Veterinary Medicine, University of Edinburgh, December 1997, 27, 33–36.

————."Well Trained Oxen," *Rural Heritage*. 1994 Summer, 19:3, 30–31.

————."Why Oxen," *Rural Heritage*. 2004 Holiday, 29:6, 83–85.

————."Working a Single Ox," *Rural Heritage*. 1998 Winter, 23:1, 26–28.

————."Working Oxen in Public," *Rural Heritage*. 1994 Holiday, 19:5, 10–13.

————."Worldwide Animal Traction," *Rural Heritage*. 1997 Winter, 22:1, 76–77.

Conroy, Drew, Claire Handy, and Lobulu Sakita. "The Adoption and Use of Animal Traction by the Agropastoral Maasai and Arusha People in Northeastern Tanzania," *Draught Animal News*. Edinburgh: Centre for Tropical Veterinary Medicine, University of Edinburgh, June 2004, 40, 2–17.

Conroy, Drew and Dwight Barney. *The Oxen Handbook*. Butler Publishing and Tools: LaPorte, CO, 1986.

Cornelius, Evelyn D. "Georgia's Lep and Lem," *Rural Heritage*. 1996 Autumn, 21:5, 72.

————."Lady Drover," *Rural Heritage*. 1997 Winter, 22:1, 30–31.

Crossley, Peter and John Kilgour. *Small Farm Mechanization for Developing Countries*. New York: John Wiley and Son, 61–71, 1983.

Dearborn, Jeff. *Cattle Hoofcare*. VHS. LaPorte, CO: Butler Publishing and Tools, 1989.

Food and Agriculture Organization of the United Nations. *The Employment of Draught Animals in Agriculture*. 1968, 32–39.

Fussner, Warren and Susan Salterberg. "Oxen – Part of Our Heritage," *The Evener*. 1984 Summer, 20–21.

Greenhall, Susan. "Distance pull for Oxen," *Rural Heritage*. 1995 Summer, 20:4, 64–65.

Grossetete, Jean Christophe. "Training Oxen in France," *Small Farmer's Journal*. 1991 Winter, 15:1, 20–22.

Hopfen, H.J. *Farm Implements for Arid and Tropical Regions*. Rome: Food and Agriculture Organization of the United Nations, 1969.

Hubbard, Chase. "Missouri Town Oxen Workshop," *Rural Heritage*. 1996 Winter, 21:1, 20–22.

Jaeger, William K. Agricultural Mechanization: The Economics of Animal Draft Power in West Africa. Boulder, CO: Westview Press, 1986.

Jainuden, M.R. "Reproduction in Draught Animals: Does Work Affect Female Fertility?" in J.W. Copland's *Draught Animal Power for Production: ACIAR Proceedings Series*. Vol. 10, 130–133, 1985.

James, Terry and Frances Alderson. "In Praise of Oxen," *Rural Heritage*. 1994 Spring, 19:1, 7–9.

Johnson, Jacklyn. "Canadian Head Yokes," *Rural Heritage*. 1997 Summer, 22:4, 20–22.

Kehoe, Michael M. and Chan Lai Chu. "Training Draught Buffaloes," *Agriculture International*. 1985 October, 222–227.

Kramer, Craig. "Ox Training," A memo regarding Peace Corps work in Togo, West Africa for *Tillers International*. Yokes and Harnessing Systems, 6–7, Hitches 978–113, 1983.

Kramer, Dave and Drew Conroy. *Ox Yokes I: Carving a Yoke*. VHS. Kalamazoo, MI: *Tillers International*, 1998.

Lawrence, M.R. "A Review of the Nutrient Requirements of Draught Oxen." In J.W. Copland's *Draught Animal Power for Production: ACIAR Proceedings Series*. Vol. 10, 84–89, 1985.

Ludwig, Ray. *The Pride and Joy of Working Cattle*. Westhampton, MA: Pine Island Press, 1995.

Marcy, Randolph B. *The Prairie Traveler: The 1859 Handbook for Westbound Pioneers*. Originally published by the U.S. War Dept. Mineola, NY: Reprinted by Dover, 2006.

Miller, Ralph C. "Oregon Trail Ox Teams," *Small Farmer's Journal*. 1977, 1:4, 12–15.

————."Oxen," *Small Farmer's Journal*. 1977, 1:4, 10–12.

Minhorst, Rolf. "The Evolution of the Draught Cattle Harness in Germany," *Small Farmer's Journal*. 1991, 15:1, 37–56.

———.*Modern Harness for Working Cattle*. Osnabruck, Germany: self published at EG Hochschulburo Weser-Ems, 1991.

Palmer, Rachel and Ben, The Oxmasters, Draught Oxen for Demonstration and Hire, Powys, England. Personal communication, 1995.

Petheram, P.J., M.R. Goe, and Abiye Astatke. *Approaches to Research on Draught Animal Power in Indonesia, Ethiopia and Australia*. Townsville, Australia: James Cook University, 1989.

Pike, Robert. "The Teamsters and The Teams" in *Tall Trees, Tough Men*. New York: W.W. Norton, 116–129, 1984.

Pingali, Prabhu, Yves Bigot, and Hans P. Binswanger. *Agricultural Mechanization and the Evolution of Farming Systems in Sub Saharan Africa*. Baltimore, Maryland: Johns Hopkins University Press, 1987.

Porter, Robert. "The Fit of An Ox Yoke," *Small Farmer's Journal*. 1985, 9:3, 26–30.

Powell, Richard. "A History of the Ox Harness," *Small Farmer's Journal*. 1989, 13:4, 46–50.

Robinson, Betty. "Arkansas Oxen," *Rural Heritage*, 1996 Summer, 21:4, 20–22.

Roosenberg, Richard. "Animal Driven Shaft Power," in *Tillers TechGuide*. Kalamazoo, MI: Tillers International, 1986.

———."Fitting Ox Bows," *Rural Heritage*. 1995 Evener, 20:3, 56–57.

———."The Single Yoke," *Rural Heritage*. 1998 Winter, 23:1, 29.

Sheldon, Asa. *Yankee Drover*. Originally published in 1862 as *Life of Asa Sheldon, Wilmington Farmer*. Reprinted by University Press of New England, 1988.

Starkey, Paul. Harnessing and Implements for Animal Traction. London: IT Publications, 1989.

Starkey, Paul and Adama Faye, eds. Animal Traction for Development: Proceedings of the 1988 West Africa Animal Traction Network. Netherlands: CTA Ede-Wageningen, 1990.

Starkey, Paul, Emmanuel Mwenya, and John Stares, eds. *Improving Animal Power Technology: Proceedings of the 1992 ATNESA Workshop*. Netherlands: CTA Wageningen, 1994.

Starkey, Paul, Sirak Teklu and Michael R. Goe. *Animal Traction: An Annotated Bibliographic Database*. Addis Ababa, Ethiopia: International Livestock Research Centre for Africa, 1991.

Suits-Smith, Kathy. "Head Yokes," *Rural Heritage*. 1997 Summer, 22:4, 18–19.

———."Nice 'n Easy Ox Teamster," *Rural Heritage*. 1997 Holiday, 22:6, 32–34.

Tillers TechGuide. "Neck Yoke Design and Fit: Ideas from Dropped Hitch Point Traditions," Kalamazoo, MI: Tillers International, 1992.

Watson, Peter. *Animal Traction: Peace Corps Manual*. Summit, NJ: Artisan Publications, 1981.

Welsch, Jochen. "Defending Oxen: A Reassessment of Their Role in American Agriculture," Pittsboro, NC: American Minor Breeds Conservancy, 1988.

———." 'If the Worcester Boys Want to See Cattle Haul, They Must Come to Kennebec': A Trial of Working Oxen as an Expression of Regional Agricultural Values 1818-1860," in *Proceedings of the Dublin Seminar: New England Creatures, 1400–1900*. Boston University Publications, 1993.

———." 'A Real Yankee Always likes to See a Good Pair of Oxen Pull': The Development and Reemergence of a Regional Agricultural Tradition." Masters of Arts Degree Thesis. Chapel Hill, NC: University of North Carolina, 1994.

———."Tarheel Ox Teamsters," *Rural Heritage*. 1995 Holiday, 20:6, 48–50.

West, Frank. *Manual on Training Steers for Oxen*. Rogue River, OR: self published Mimeograph, 1986.

Weston, R.H. "Some Considerations of Voluntary Feed Consumption and Digestion in Relation to Work." in J.W. Copland's *Draught Animal Power for Production: ACIAR Proceedings Series*. Vol. 10, 78–83, 1985.

Wheelock, Janelle. "Old Fashioned Oxen Adapt Well to the Modern Small Farm," *Countryside and Small Stock Journal*. 1992 Nov-Dec, 76:6, 28–29.

Youatt, William. *Cattle*. London: Baldwin and Cradock, 1842.

Index

Page numbers in **bold** indicate tables.

BRSV (bovine respiratory syncytial virus), 210
brucellosis, 211
building roads with oxen, 154–56
"bull punchers," 144–45, 176
"bull whackers," 144
Butler, Doug, 77
BVD (bovine virus diarrhea), 210

C

calabash (grab) ring, 135, 136, 137
calcium-to-phorphorus ratio, 48
calves
    feeding, 39–40, 41, 42, 53, 54
    selecting, 6, 8, 20
    training principles, 56, 58, 59, 60, 62, 186–87
    training steers, 78–92
canarygrass, **276**
capture and restraint, training mature cattle, 104
carbohydrates, 39, 41
caring for yokes and bows, 134, 138
carts and wagons, 169–70
castration impact, 2
Castro, Fidel, 109
chain, practicing pulling with, 86, 88, 94
chain length, 98
challenges, introducing to cattle, 82–83, 90
challenges of international development, 263–64
changing diet gradually, 217, 218, 219
Charolais
    breed comparison, 12, **270–71**
    examples, 9, 10, 13, 22, 77, 123
chemical sprayers with oxen, 172, 174
cherry *(Prunus serotina)*, 126
Chianina
    breed comparison, 12–13, **270–71**
    × Charolais, 249
    examples, 5, 10, 11, 22, 32, 42, 101, 204, 235
    × Holstein, 20, 23, **270–71**
children caution, working oxen in public, 188, 189
choosing a breed, 9–10. *See also specific breeds*
chutes, restraint, 224–25, 226, 227
"circling disease," 211
cleanliness and sanitation, 210
clearing land with oxen, 152–54
clinching shoe, 233–34, 236
clipping (body-clipping), 199
clover, **276**

Coccidia, 213, 215
Collins, Don (Berwick, ME), 107, 221
Colonial Williamsburg Foundation (CWF), 259
color of animal and selecting the ideal team, 8
colostrum, importance of, 39–40, 213
Columella, 252
"come haw" (turn left), 79, 85, 87–88, 89
comfort before beauty, making neck yokes and bows, 130
commands, teaching, 79, 84–90. *See also* training
commercial protein mix, 47
common breeds, 10–18
comparison of breeds, **270–71**
competing with oxen, 193–206
competitions (local) for international development, 268
concentrates (grains), 39, 42, 43, 44, 46–47, 51, 52, **273**, 275, **276, 277**
concrete flooring, 32–33
conditioning, 94–96, 156–57, 159, 160
conformation, selecting the ideal team, 3, 4–5, 8, 20, 225–27
conformation classes, 194, 195–96
Conroy, Drew, 77, 269
Conroy, Janet (Berwick, ME), 35
Conroy, Ross and Luke (Berwick, ME), 39
consistency for training, 81, 90
contouring a field with oxen, 159
control (maintaining), 80–81, 186
corn, 45, 46, **276**
coronavirus, 210
cottonwood *(Populus tremuloids, P. grandidentata)*, 126
"cowboy style" of training, 107
cow-hocked, 5
cows for work and reproduction, feeding, 50
Cowslips, 236
CP (crude protein), 44, **273,** 274, **276–77**
cracks in yokes, avoiding, 126–27
creativity for international development, 263–64
Criollo cattle, 18
crossbreeds, common, 11, 20
Crucible, The (film), 221, 257
crude protein (CP), 44, **273,** 274, **276–77**
cud-chewing, 209
cultivating with oxen, 161–62
cultural differences in competitions, 202

cutting the load, 96

D

dairy cattle, 9, 10, 42
Dairy Cattle Science (Ensminger), 39
Daly, Kevin (Lyndonville, VT), 249
designing neck yoke and bows, 124–39
developing countries, 52, 251–52, 261–62
Devon (American Milking Devon, Milking Devon)
    × Ayrshire, 77
    breed comparison, 13–14, **270–71**
    examples, 7, 9, 10, 11, 23, 29, 41, 53, 61, 65, 77, 120–21, 205, 221, 255, 256, 257
    × Holstein, 20, 24, 77, 249, **270–71**
    × Lineback, 205
    × Normandy, 77
deworming, 42, 213, 214, 215
Dexter
    breed comparison, 14, **270–71**
    examples, 10, 24, 32
diarrhea (scours), 40, 212, 213
diet. *See* feeding oxen
differences among breeds, 3, 7
digestive tract of bovine, 38–39, 217–18, 219–20
disadvantages, training mature cattle, 103
disc harrow, 161, 163
discomfort of yoke, signs of, 115, 119, 138, 220, 221
disease symptoms, awareness of, 209
displaced abomasum, 219
disposition of animal and selecting the ideal team, 8, 10, 20
distance pull contests, 203
DMI (dry matter intake), **272, 273,** 275, **276–77**
domestication of cattle, 251
dominance
    cruelty vs., 79
    training and, 56–57, 58–60, 62, 68, 77, 81, 90, 107–8, 245
draft animal power (animal traction) programs, 267–68, 269
draft horses vs. oxen, 149, 174
draft-horse technology and oxen, 151, 174
drilling bow holes, 129
driver's weight in mower, 164
driving a multiple hitch, 145
driving from behind, 99
drop chains for multiple teams, 146
dropped hitch point, 118, 142

dry matter intake (DMI), **272, 273,**
255, **276–77**
dry vs. wet environment, 32, 36
Durham cattle. *See* Milking Shorthorn
Dutch Belted (Lakenvelders,
Lakenfeld Cattle)
breed comparison, 14–15, **270–71**
examples, 10, 24, 107, 189–90, 191,
212

E
early history of oxen, 251–52
East African Short Horned Zebu, 19,
115
Eastern Europe, 254, 261–62
educating the public, 189
educational approaches for international
development, 267–68
electric fences, 36
elimination pull contests, 203–4
elm *(Ulmus americana),* 126, 139
ending training on a positive note, 83
energy, feed, 39, 41, 42, 43, 44, **272**
Ensminger, M. E., 39
environment impact on
feeding oxen, 51
selecting the ideal team, 7, 9–10, 19,
20
training, 60
equipment for pulling, 94
Escherichia coli, 213
European explorers and oxen, 253–54
event (knowing your), working oxen in
public, 189–90
expecting the unexpected, working
oxen in public, 188, 192
experienced team, using for training
mature cattle, 107
exposure to disease, minimizing, 208
external parasites, 215–17

F
face net, 217
farmer to farmer approach for
international development, 267
farming. *See* agriculture, oxen in
farm team, selecting the ideal team,
3, 6
fat ox classes, 197
fats, feed, 39
feeders, 34, 52
feeding oxen, 37–54, **276–77**. *See also*
health care
housing and, 30, 33, 34, 46
fences, 36
fertilizer spreaders with oxen, 172

fescue, **277**
fiber, feed, 44
fiberglass yokes, 130–31
firewood hauling with oxen, 178, 179,
182
first steps, training steers, 79
fit of yoke, importance of, 111, 119–20,
138, 220, 221, 222, 244, 246, 265
fitting and showmanship (4-H) classes,
199–200
fitting the shoe, 232
flies, 216–17
flight distance of animal, 61–62, 103
flooring, housing, 32–33
flukes, 214
footrot, 211, 226
forages (hay, silage, pasture), 39, 41,
42, 43, 44–46, **276, 277**
forecarts, 151, 182
forest thinning operations with oxen,
178
4-H and youth shows, 195, 196, 197–
200, 205, 206
free-choice forage, 39, 41
free-for-all pull contests, 203, 204
Freeman, Lee (Thornton, NH), 53
Fulani, 19
fungi, 216

G
gastrointestinal obstructions, 219
"gee," "gee off," "back gee" (turn
right), 85, 86–87, 88
Gelbvieh, 19
genetic factors, impact on training, 60
geographical considerations, selecting
the ideal team, 7, 9–10, 19, 20
"get up" command, 106
Giardia spp., 214
global breeds, 19
global use of oxen, 149–50, 251–
52, 258. *See also* international
development
Gloucester cattle, 17
goad stick for training, 85, 86, 87, 88,
89
goals
selecting the ideal team and, 3, 6, 20
setting for training, 80, 90
grab (calabash) ring, 135, 136, 137
grains (concentrates), 39, 42, 43, 44,
46–47, 51, 52, **273,** 275, **276,
277**
grass hay and silage, 45
grazing in the yoke (problem and
solution), 247–48

Greece, 251
grooming
competitions and, 199–200
working oxen in public and, 190
ground skidding, 170
growing steer, feeding, 41–42
growth, monitoring, 41, 43
grubs, 214, 215
Guernsey, 10, 19, 42, **270–71**

H
habituation, 66
Haemophilus somnus, 212
halter training, 104, 105–6
handling animals carefully and safely,
10, 57, 58, 59, 60, 62, 67, 76
hardware disease, 219–20
hardware for neck yoke and bows,
125, 135–38
hardwoods for neck yoke and bows, 126
harness, 112. *See also* yoke, styles
*Harnessing and Implements for Animal
Traction* (Starkey), 159–60
harrowing with oxen, 160–61
harvesting forage crops with oxen, 164
hauling out (problem and solution),
241–42, 249
hauling stones with oxen, 154
"haw," "come haw," "back haw" (turn
left), 79, 85, 87–88, 89
hay
feed, 43, 44, 45, 50, 51, 52, **273,**
275, **276, 277**
harvesting with oxen, 164–68
head-shy caution, 106
head yokes, 112–14, 116, 123, 253
health care, 207–22
feeding oxen, 37–54
hoof care, 223–36
health issues
feeding oxen, 45–46, 48, 49–50, 51,
52
housing oxen, 34, 36
healthy hoof, 226
heat stress caution, 95, 160, 165, 189
heel flies (warble), 214
herd animals, cattle as, 57, 58, 59, 144
Hereford (Polled Hereford, Horned
Hereford)
breed comparison, 15, **270–71**
examples, 9, 10, 25, 42, 254, 257
× Holstein, 25, 77, **270–71**
Highland, 26
Hine, Art, 123, 191
Hine, Nathan, Duane, and Russell
(Worthington, MA), 123

history, oxen in, 250–59
  agriculture, 148–74
  logging, 176, 177
hitching
  multiple teams, 145–46
  options, 140–47
  single ox, 11, 141–44, 146
  training, 72–73
hitch point, 115, 118, 142, 143
Holstein
  breed comparison, 16, **270–71**
  × Chianina, 20, 77
  × Devon, 20
  examples, 9, 10, 12, 13, 15, 20, 25,
    42, 77, 101, 107, 109, 191
  × Hereford, 77
  × Shorthorn, 20, 257
hoof
  care, 223–36
  conformation, 225–27
  injuries, 222
Horned Hereford. *See* Hereford
horn fly, 217
horns caution, 57, 64
horse harness, 112, 143
Horses, Oxen and Technological Innovation
  (Langdon), 253
horses vs. oxen, 149, 160, 174, 176,
  178, 183, 251, 255
housing oxen, 30–36
humped breeds *(Bos indicus)*, 2, 10, 19,
  28, 115, 265
Huppe, Tim (Farmington, NH), 53,
  139, 259
hybrid vigor, 11

I
IBR (infectious bovine rhinotracheitis),
  210
ideal team, selecting the, 1–29. *See also*
  breeds of oxen
imitation, 66
imprinting, 65–66, 67
inattentiveness (problem and solution),
  245–46
individuality of animal and training,
  60, 61, 79–80
infectious bovine rhinotracheitis
  (IBR), 210
infectious diseases, 209–12, 222
initial acclimation, training mature
  cattle, 103
initial training, training mature cattle,
  104–6
injury and trauma, 220, 222
insurance for working oxen in public, 192

intelligence of cattle, 56, 65–68
internal parasites, 213–15
international development, 260–69.
    See also agriculture, oxen in
intramuscular shots, 209
introducing oxen for international
  development, 264–65

J
James, Dave, 176
Jersey
    breed comparison, 17, **270–71**
    examples, 10, 16, 27, 42
Jersey × Holstein, 20, *27*, **270–71**
joint ill (navel infection), 212
judging competitions, 195, 198, 200, 201
jumping chain (problem and solution),
  240–41, 249

K
Kentucky bluegrass, **277**
kicking (problem and solution), 244
Kluchnik, Bud (Ripley, ME), 91

L
Lakenvelders, Lakenfeld Cattle. *See*
  Dutch Belted
Lamb, Irving ("Johnny"), 77
lameness, 49–50, 222
laminated wood yokes, 131
laminitis, 45, 218–19, 226
land (clearing) with oxen, 152–54
Langdon, John, 253
Latin America, 261
leaders (natural), 59
leading cattle, 75
leading with halter, 82
learning by cattle, 65–68
learning from successful teamsters, 56
left (nigh) side of yoke, 80
leg
    conformation, 225–27
    injuries, 221, 222
legume hay and silage, 45
leptospirosis, 211
Lester, Gary, 77
lice, 215
life span of oxen, 7, 50, 54, 197, 206
limit-feeding, 33, 48–49
Limousin x Holstein, 183
Lineback
    breed comparison, 17, **270–71**
    examples, 16, 27, 77
listeriosis, 211–12
load guidelines, 93–94, 95, 96
loading hay with oxen, 167

local competitions for international
  development, 268
lock jaw (tetanus), 213
logging contests, 194, 195
logging lizards, 180
logging wheels (big wheels), 182
logging with oxen, 68, 152, 175–83
log scoot contests, 201
Longhorn. *See* Texas Longhorn
loose housing, 31, 32
Ludwig, Ray, 77, 166
lumpy jaw, 212
"lunge space," 32
lungworms, 214
lying down in the yoke (problem and
  solution), 243–44
Lyme disease, 216

M
Maasai farmers, 251–52, 266, 269
*Maasai Oxen, Agriculture and Land Use
    Change* (Conroy), 173
machinery vs. logging with oxen, 177,
  178
Madagascar Zebu, 19
maggots and screwworms, 216
maintaining control, 80–81, 186
making neck yoke and bows, 124–39
Manning, J. Russell, 9
manure spreaders with oxen, 171, 174
maple *(Acer rubrum)*, 126, 139
marketing tool, oxen as, 172
Marston, Lester, 205
matching ("mating") cows, selecting
    the ideal team, 3
mature oxen
    feeding, 48–50, 54
    selecting, 7, 20
    training, 102–9
Mcal/lb (megacalories) per pound of
    feed, 44
meadow hay, **277**
media for international development,
  268
megacalories (Mcal/lb) per pound of
    feed, 44
memory of painful or disagreeable
    experiences, 186
metabolic diseases, 217–20
Mewati, 19
Middle Ages and oxen, 253
Midwest Ox Drovers Gathering, 191
Milking Devon. *See* Devon
Milking Shorthorn (Shorthorn,
    Durham cattle)
    breed comparison, 17–18, **270–71**

Other Storey Titles You Will Enjoy

Barnyard in Your Backyard, edited by Gail Damerow.
Expert advice on raising healthy, happy, productive farm animals.
416 pages. Paper. ISBN 978-1-58017-456-5.

Building Small Barns, Sheds & Shelters, by Monte Burch.
Comprehensive coverage on tools, materials, foundations, framing, sheathing,
wiring, plumbing, and finish work for outbuildings.
248 pages. Paper. ISBN 978-0-88266-245-9.

Getting Started with Beef & Dairy Cattle, by Heather Smith Thomas.
The first-time farmer's guide to the basics of raising a small herd of cattle.
288 pages. Paper. ISBN 978-1-58017-596-8.
Hardcover with jacket. ISBN 978-1-58017-604-0.

Grass-Fed Cattle, by Julius Ruechel.
The first complete manual in raising, caring for, and marketing grass-fed cattle.
384 pages. Paper. ISBN 978-1-58017-605-7.

How to Build Animal Housing, by Carol Ekarius.
An all-inclusive guide to building shelters that meet animals' individual needs:
barns, windbreaks, and shade structures, plus watering systems, feeders, chutes,
stanchions, and more.
272 pages. Paper. ISBN 978-1-58017-527-2.

How to Build Small Barns & Outbuildings, by Monte Burch.
Complete plans and instructions for more than 20 projects, including an add-on
garage, a home office, a roadside stand, equipment sheds, and four types of barns.
288 pages. Paper. ISBN 978-0-88266-773-7.

Keeping Livestock Healthy, by N. Bruce Haynes, DVM.
A complete guide to disease prevention through good nutrition, proper housing,
and appropriate care.
352 pages. Paper. ISBN 978-1-58017-435-0.

Livestock Guardians, by Janet Vorwald Dohner.
Essential information on using dogs, donkeys, and llamas as a highly effective,
low cost, and nonlethal method to protect livestock and their owners.
240 pages. Paper. ISBN 978-1-58017-695-8.
Hardcover. ISBN 978-1-58017-696-5.

Small-Scale Livestock Farming, by Carol Ekarius.
A natural, organic approach to livestock management to produce healthier
animals, reduce feed and health care costs, and maximize profit.
224 pages. Paper. ISBN 978-1-58017-162-5.

These and other books from Storey Publishing are available
wherever quality books are sold or by calling 1-800-441-5700.
Visit us at *www.storey.com.*